AFTER ATHEISM

CAUCASUS WORLD
SERIES EDITOR NICHOLAS AWDE
www.caucasusworld.co.uk

Other books in the series include:

Small Nations & Great Powers: A Study of Ethnopolitical Conflict in the Caucasus *Svante Cornell*
The Russian Conquest of the Caucasus *J. F. Baddeley — with a new Preface by Moshe Gammer*
Storm Over the Caucasus: In the Wake of Independence *Charles van der Leeuw*
Oil & Gas in the Caucasus & Caspian: A History *Charles van der Leeuw*
Daghestan: Tradition & Survival *Robert Chenciner*
Madder Red: A History of Luxury & Trade *Robert Chenciner*
Azerbaijan: Quest for Identity — A Short History *Charles van der Leeuw*
The Georgian-Abkhaz War *Viacheslav A. Chirikba*
Georgia: In the Mountains of Poetry *Peter Nasmyth*
The Literature of Georgia: A History *(2nd, revised edition) Donald Rayfield*
The Armenian Kingdom in Cilicia during the Crusades: The Integration of Cilician Armenians with the Latins, 1080-1393 *Jacob Ghazarian*
A Bibliography of Articles on Armenian Studies in Western Journals, 1869-1995 *V. N. Nersessian*
Armenian Perspectives *edited by Nicholas Awde*
Armenian Sacred & Folk Music *Komitas (Soghomon Soghomonian), translated by Edward Gulbekian*
The Armenian Neume System of Notation *R. A. Atayan, translated by V. N. Nersessian*
Ancient Christianity in the Caucasus (Iberica Caucasica vol. 1) *edited by Tamila Mgaloblishvili*
The Cross and the Cresecent: Early Christianity & Islam in the Caucasus (Iberica Caucasica vol. 2) *edited by Tamila Mgaloblishvili (forthcoming)*
Monasticism in the Early Christian East and West (Iberica Caucasica vol. 3) *edited by Tamila Mgaloblishvili (forthcoming)*
Pilgrimage: Timothy Gabashvili's Travels to Mount Athos, Constantinople & Jerusalem, 1755-1759 *edited by Mzia Ebanoidze & John Wilkinson*
Also available for first time in paperback: The Man in the Panther's Skin *Shot'ha Rust'haveli (Shota Rustaveli), translated by Marjory Scott Wardrop* (ROYAL ASIATIC SOCIETY)

PEOPLES OF THE CAUCASUS & THE BLACK SEA

1. The Armenians
2. The Georgians
3. The Azerbaijanis
4. The Chechens
5. The Abkhazians
6. The Circassians
7. The Peoples of Daghestan
8. The Ossetes
9. The Ingush
10. The Turkic Peoples of the Caucasus
11. The Iranian Peoples of the Caucasus
12. The Mountain Jews
13. The Georgian Jews
14. The Laz
15. The Mingrelians & Svans
16. The Ubykh
17. The Displaced Peoples of the Caucasus in Soviet Times
18. The Caucasus in Diaspora
19. The Hemshin
20. The Kalmyks *edited by Elza-Bair Gouchinova & David C. Lewis*
21. The Cossacks
22. The Ancient Peoples of the Caucasus
23. The Crimean Tatars
24. The Gagauz
25. The Karaim
26. The Pontic Greeks

CAUCASUS LANGUAGES

Chechen Dictionary & Phrasebook *Nicholas Awde & Muhammad Galaev*
Georgian Dictionary & Phrasebook *Nicholas Awde & Thea Khitarishvili*
Armenian Dictionary & Phrasebook *Nicholas Awde, V. Nersessian & L. Nersessian*
Azerbaijani Dictionary & Phrasebook *Nicholas Awde & Famil Ismailov*
The Languages of the Caucasus *edited by Alice Harris & Rieks Smeets [forthcoming]*

Previous page: A ritual among the Mansi people of north-west Siberia involving the sacrifice of a pony to the spirits

AFTER ATHEISM
Religion and Ethnicity in Russia and Central Asia

David C. Lewis

CURZON
CAUCASUS WORLD

CAUCASUS WORLD

First published in 2000
by CURZON PRESS
15 The Quadrant, Richmond
Surrey TW9 1BP
England

© David C. Lewis 2000

Typeset and designed by Nicholas Awde/Desert♥Hearts
Scans by Emanuela Losi & David McDonald
Covers & maps by Nick Awde
Photos by David C. Lewis unless otherwise credited

Printed and bound in Great Britain by
Bookcraft, Midsomer Norton, Avon

All rights reserved. No part of this book may be reprinted or reproduced or utilised in any form or by electronic, mechanical, or other means, now known or hereafter invented, including photocopying and recording, or in any information storage or retrieval system, without permission in writing from the publishers.

British Library Cataloguing in Publication Data
A catalogue record for this book is available from the British Library

ISBN 0 7007 1164 3

Contents

	Map of places visited	6
	Map of ethnic groups	8
	Map of religions	10
	Acknowledgements	11
	Introduction: Spiritual awareness	13
1.	Peoples and cultures	23
2.	Dreams and visions	52
3.	Fear	73
4.	Death	96
5.	Who am I?	124
6.	Finding one's roots	142
7.	The Orthodox kaleidoscope	177
8.	Free market religion	202
9.	Human rights	223
10.	Compassion	240
11.	Miracles	263
12.	Facing the future	275
	Notes	299
	Bibliography	311
	Index	317

After Atheism

Map of places visited

After Atheism

Map of ethnic groups

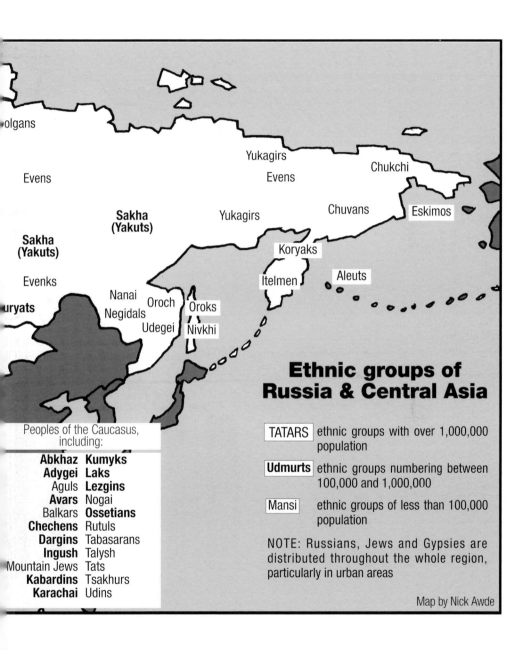

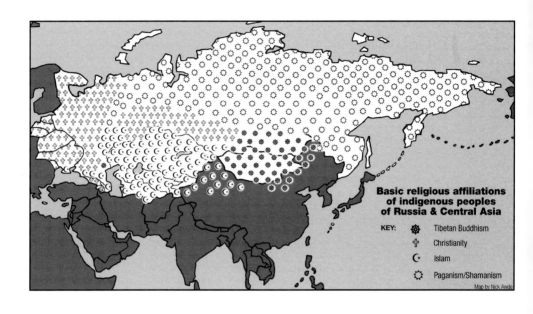

Acknowledgements

To a large extent this book draws upon my own interviews and first-hand ethnographic fieldwork in several different parts of Russia and Central Asia. At the same time I am greatly indebted to many scholars and other individuals who have helped me by their comments, information and advice. At academic seminars or conferences in Russia, Central Asia or elsewhere I have benefited from discussion with a variety of scholars but I have to admit that sometimes I have forgotten the names of those who made comments or suggestions. However, among those whose help stands out in my mind can be included the following:

- *in Udmurtia:* Marina Khodyreva, Alexander Ivanov, Lydia Orekhova, Vladimir and Tatyana Vladykin, Mikhail and Granya Yegorov, Mariya Slesareva, Lutsiya Volkova, Pavel Zhelnovakov, Irina Voronchikhina, Nadezhda Zolotareva, Boris Karakupov, Irina Khodyreva, Irina Nureyeva;
- *in Tatarstan:* Shamil Fattakhov, Farida Sharifullina, Nuriya Garayeva, Sergei and Julia Borisenkov, Yuri Mikhailov, Gali Maksumov, Kurbangali and Nadia Sharipov, Farida Zhestovskaya, Lilya Korchagina, Yevgeny Aronson, Lilya Latfullina;
- *in Bashkortostan:* Rail Kuzeyev, Irina Tagirova, Boris Razveyev, Vladimir Silchuk, Gulnara Abdurashitova, Fanis Gabidullin, Ilgiz Safarov, Yuri Afanasyev, Idus Ilishev;
- *in the Komi Republic:* Svetlana Nizovtseva, Yuri Shabayev, Daniel Popov, Oleg and Irina Artemyev;
- *in the Mari-El Republic:* Valeriy Patrushev; Vladimir Sharov, Neilya Minyazhetdinova;
- *in Kalmykia:* Elza-Bair Gouchinova, Emma Gabunshina, Larissa Jabrueva, Sarang Badeyev, Sergei and Nadia Sychov, Gilyana Orgadulova;
- *in the Khanty-Mansi okrug:* Gennady and Valentina Solovar, Alexander and Lena Fedotov, Andrei Golovnev, Yuri Vella, Valentina Litvinenko, Zinaida Katanina, Lyubov Lushchai, Vera Lebedeva and many others in the villages of Lombovozh, Sosva and Kimkyasui;
- *in the Sakha Republic:* Claudia Fyodorova, Alexandra Petrova, Alexander Sharin, Yeremy Tabishev, Vadim Naumov, Olga Parfenova, Radomir and Larissa Borisov, A. N. Alexeyev, Nikolai Kurilov, the family of Vladimir Kurilov, Innokenty Trityakov, Nikolai Neusterov, Polina Kurilova and many others in the village of Andryushkino;

*in **Yekaterinburg:*** Lyudmila Ivanko, Lena Glavatskaya, Pavel Bak, Tanya Osintseva, Vladimir and Farida Lebedev, Tanya Govorukhina, Boris Bagirov, Vera Panova, Dmitry Abramov, Benjamin Golubitskiy, Andrei Ivanov;

*in **Moscow:*** Vladimir and Irina Basilov, Sergei Arutiunov, Andrei and Karina Chernyak, Lawrence Uzzell, Lyudmila Nikishyova, Yuri Kulchik, Rosa Jarylgasinova, Olga Ostrovskaya, Irina Galoyan, Vladimir and Nadia Katkov, Maxim Shub, Olga Pakhotina;

*in **St Petersburg:*** Vladimir and Olga Bronnikov, Rakhmat Rakhimov, Alexei Konovalov, Olga Khazhnyak;

*in **Kirov:*** Irina Trushkova, Natalya Zaitseva;

*in **Novosibirsk:*** Yelena Fursova;

*in **Astrakhan:*** Victor Viktorin;

*in **Chukotka:*** Nuriya Yusupova, Alexei Astapov, Igor Keleveket, Mikhail Badanov and many others in the village of Lorino;

*in **Kazakstan:*** Nurilya Shakhanova, Karlygash Ergazieva, Asya Kamalidenova, Marat Khasanov, Alexei Kopeliovich, Raushan Mustafina;

*in **Turkmenistan:*** Adjap Bairieva, Guncha Yolbarsova, Keyik Yolbarsova;

*in **Armenia:*** Khachik Stamboltsyan, Gayane Marikian, Ashot Stepanian, Gohar Marikian, Manya Kazaryan, Levon Bardakjian;

*in **Cambridge (England):*** Caroline Humphrey and many others at the Mongolia and Inner Asia Studies Unit, University of Cambridge.

I am most grateful for all the help and information given by these and many others in the course of my research. Knowledge, however, also entails responsibility, and therefore I have wanted to do what I could to help alleviate some of the human suffering which I have encountered on my visits to parts of the former USSR. Often I have brought with me humanitarian aid of various kinds (such as medicines, children's clothes, toys and so on) and have been grateful for the help of SAS (Scandinavian Air System) since 1994, and of British Airways from 1993 to 1994, in transporting such materials as far as Moscow. From there onwards numerous local people have not only helped me physically (in terms of transport, hospitality or accommodation) but have also shared with me their inner thoughts and feelings. Included among these are their spiritual experiences. Often I have quoted these anonymously but I want to take this opportunity to express my gratitude to my many friends whose lives and experiences are reflected in the pages of this book.

Introduction:
Spiritual awareness

"Akhtyam! Akhtyam!" As Akhtyam awoke from sleep, he could hear clearly his mother's voice calling to him. He had not been dreaming of her, but he could hear her voice calling to him as he woke up. However, she was 300km away in the city of Kazan.[1] That same evening, on 21st August 1997, Akhtyam received a telegram from Kazan informing him that his mother was seriously ill and was partially paralysed after having had a stroke. He travelled to Kazan as soon as he could do so. Each weekend after that he returned to Kazan from his home in the town of Almetevsk, in south-eastern Tatarstan. It was on one of those journeys, on Friday 26th September, that he told me about his experience of hearing his mother's voice just one month previously.

Prior to that time, Akhtyam had regarded himself as an atheist. Now he did not know what to believe. Certainly his experience could not be discounted, and it made him believe now that there is some kind of spiritual or supernatural dimension to life. Was his experience some kind of psychic phenomenon like telepathy? Or might it have been a revelation from God? If so, was this from Allah — as Akhtyam belongs to the nominally Muslim Tatar people — or from some other source?

Experiences of this kind challenge our preconceived ideas and our conventional worldviews. For those brought up in an atheistic environment, such experiences can pose questions for which there are few easy answers. Sometimes they can talk about their experiences with relatives or friends, but there can also be occasions when people are afraid to speak about what has happened for fear of being ridiculed or even thought to be mentally unbalanced. Such fear was probably more pronounced during the Communist period when people were officially supposed to be 'atheists'. It was simpler to keep silent or else to confide in only one or two trusted people.

Even today, whom can one trust to take one seriously if one has an important spiritual experience? For instance, in January 1997 a friend of mine named Irina telephoned me and said that she now believed there is a God. Twelve days previously she had seen an angel. Irina had become conscious of God's care and love for her, because the angel appeared to be standing over her and seemed to be guarding her. At that time she had been in bed with a severe bout of flu and a temperature of 40°.

We might ask whether this might have been a hallucination or whether it was a genuine vision of an angel. Irina herself had no doubt that hers was a genuine and very meaningful spiritual experience. However, I suspected that she decided to telephone me in England to talk about it because she felt that I would be more likely than some of her other friends to take her seriously and not to laugh at her.[2]

From 1985 to 1988 I had worked for a research centre in Oxford which conducts investigations into the incidence and nature of religious or spiritual experience.[3] The centre was founded by Sir Alister Hardy, a professor of marine biology who had become interested in the fact that many normal people have experiences of a spiritual nature. Subsequent studies in Britain, the USA and elsewhere have indicated that between a third and a half of the general population have had some kind of 'supernatural', 'religious' or 'paranormal' experience at some time in their lives.[4] In some social subgroups the proportion might be even greater, as shown by the study I conducted among a random sample of nurses which showed that two-thirds of my sample reported a definite spiritual experience of some kind or other: most of the remainder reported less definite accounts or could not remember concrete details of their experiences.[5]

Preliminary indications from my interviews with Russians, Mongolians, Tatars and other peoples of Siberia and Central Asia indicate that spiritual experiences are not uncommon among them too. However, their upbringing in an officially atheistic social environment has meant that many people are now unsure about the consequences of their experiences. For instance, Irina, mentioned above, belongs to the Bashkort people of the southern Urals region. Prior to the Communist period the Bashkorts had been at least nominally Muslim, but after some seventy years of atheism the younger generations have been brought up without a religious consciousness. Irina told me that she had previously regarded herself as an atheist, but after her vision of an angel she had come to believe there is a God. However, in her words, she did not know whether she should now become "a Buddhist or a Baptist."

What is 'religion'?

One legacy of the Communist period is that reports of spiritual experiences tended to be suppressed, probably partly out of fear and partly out of a lack of knowledge about how common such experiences actually are. Nowadays there is a greater awareness of the fact that such experiences occur, but often people do not know how to interpret them. If the experiences are expressed to others, their interpretation is likely to be affected by social and cultural presuppositions. For instance, Akhtyam could try to interpret his experience in terms of an atheistic worldview or else he could adopt a religious interpretation. It is likely that any religious point of view would be influenced by the fact that he belongs to the nominally Muslim Tatar people. However, the kind of experience reported by Akhtyam could quite conceivably have been recounted also by a

Spiritual awareness

Ukrainian with a Roman Catholic or Eastern Orthodox background, or by a Buryat person from southern Siberia with a Buddhist or shamanistic background.

Therefore it is not surprising that during the Communist period there was a widespread interest in psychic phenomena such as telepathy or clairvoyance.[6] A religiously neutral terminology could be employed which invoked the concept of a psychic energy called 'psi' which had an appearance of scientific legitimacy. This provided a politically 'safe' vocabulary which allowed investigation into this realm without the use of religious language. For atheists, no other type of explanation for their spiritual experiences was officially available.

Nevertheless, there are many people who find this level of explanation unsatisfactory. This might simply be a 'gut reaction' at an emotional rather than a rational level. For instance, recently a Finnish lady (who, as far as I could tell, had no significant religious adherence) remarked to me that she could not accept that her relationship with her boyfriend and their mutual love for one another was simply the result of an impersonal force — whether one referred to such a force as 'chance', 'fate', 'destiny' or 'karma'. Inside herself, she felt that their relationship based on love had a deeper quality to it than could be satisfactorily explained away by reference to impersonal or abstract kinds of spiritual forces. This kind of 'gut reaction' probably predisposes some people to see their lives as having meaning in terms of relationships — including not only relationships at a human level but also in some way at the level of a spiritual world inhabited by beings possessing consciousness and personality. This kind of concept has been the dominant one both throughout history and also around the world among probably all human cultures. In terms of history, geography and social anthropology, atheism is an anomaly.

Atheism has been an official teaching not only under Communism but also within the theoretical levels of Buddhist philosophy. In practice, however, ordinary people in such cultures still tend to believe in, and offer prayers to, animate spiritual beings. Not only is there often a concept of a high God, but commonly there are also beliefs in other personalities, such as angels or demons. It is well known that even those professing to be atheists often find themselves crying out to God to help them when they come face to face with trauma, danger or death. An atheistic worldview is actually quite difficult to sustain in the face of this popular tendency for ordinary people to seek meaning in terms of a religious view of the universe. It was therefore a remarkable achievement for the Communists in Russia to have managed to maintain their ideology for as long as seventy years, even despite the fact that it was done so largely through the use of fear and coercion.

There are several problems entailed in researching religion in a post-atheistic culture. Even the very subject matter itself can be problematic: what, after all, is 'religion'? Is it merely 'a belief in spiritual beings', as Tylor suggested in the 19th century?[7] One problem with such a definition is that it focuses too much on belief to the exclusion of practice, in so far as Buddhism in theory can be regarded as 'atheistic', not necessarily involving a 'belief in spiritual beings', but in practice it strongly resembles a religion and is normally regarded as one.[8] Certainly at a 'folk' level it does include many 'religious' beliefs and practices. Therefore over the past century a number of other definitions of religion have been proposed, none of which seems to be fully satisfactory. The problem is that 'religion' embraces such a wide range of practices and beliefs that one can almost always find an exception to any definition which is too narrow. On the other hand, too broad a definition allows in that which perhaps ought to be excluded.

A case in point is atheistic Marxism, as practised in the former USSR, Eastern Europe, China and elsewhere. Many of the rituals advocated by the Communist parties in these countries appear to have replaced religious practices and to have been substitutes for religious expression. For instance, the 'pilgrimage' to Lenin's tomb on Red Square might be compared with pilgrimages among Russian Orthodox or 'Old Believer' communities. Of course, at a different level of analysis, it is true that an embalmed corpse has none of the mystery and majesty surrounding the Orthodox emphasis on Easter — with a focus on Christ's empty tomb. Nevertheless, there is a strong case to be made for seeing many Communist practices as imitations of Orthodox ones. Instead of icons, there have been pictures of Lenin or Stalin; more specifically, representations of Lenin with children are reminiscent of those depicting Christ welcoming children. Communist parades and festivals celebrate the new Socialist heroes — perhaps equivalent to Orthodox saints. There is even a Communist equivalent of 'worship' in slogans such as 'Glory to the Communist party of the Soviet Union!'[9] The list could go on, but the comparisons are fairly plain.[10]

Obviously atheism — as a system which opposes the idea of any divine being — cannot be regarded as a religion in itself. Nevertheless, the actual practice of those espousing this ideology has taken on quasi-religious forms. Like matter and anti-matter, religion and atheism appear to have similar properties but opposite polarities.

However, there is far more to religion than merely practice and belief. Commonly anthropologists have focused, perhaps too narrowly, on these two obvious dimensions of religion and have overlooked at least two other major dimensions — ethics and spiritual experience. In many ways religion is multi-layered, with each level influencing each of the others. I do not propose any all-embracing definition of 'religion' because the boundaries of what constitutes 'religion' are often fluid. In Japan, Russia and Britain many people claim not to be 'religious' but in practice they subscribe at times to behaviours which to an outsider strongly appear to be 'religious'![11] What people themselves regard as 'religious' is subject to individual negotiation and subjective perceptions of what constitutes 'religion'. Instead, I shall address those dimensions of life which appear to be motivating factors in prompting people towards behaviour of a 'religious' nature.

At least three different levels of 'religiosity' might be delineated: (1) spiritual experiences such as precognitive dreams, premonitions, visions and so on, which are often unsought but which challenge previous 'atheistic' assumptions; (2) issues of life which might evoke religious questions — e.g. ethical issues, or else life crises such as bereavement which bring a person to reassess beliefs and values; (3) organised religious groups and formalised doctrines.

Often the first level involves responses which are open to different interpretations but are not so specific to particular religious groups: for instance, Akhtyam's experience might be seen by one person as a revelation from God but by another person as merely 'coincidence' or imagination. Some cultural influence can nevertheless be discerned in certain cases, to the extent that, for example, more significance is attached to dreams in some cultures than in others.

Cultural influence is more obvious at the second level, in so far as there are certain social expectations about whether or not religious specialists are consulted about ethical or personal crises.[12] Nevertheless, such crises can still motivate people to think about 'spiritual' issues even in cultures such as those dominated by atheistic ideologies where 'religious' tendencies are actively discouraged.

The third level of 'religious' behaviour — that of organised groups — is the most obvious one and is therefore the one to which most attention has been paid. Partly this is because it often involves issues of cultural or ethnic identity. How one interprets the experience often involves an interaction between the spiritual experience itself and the cultural context in which the person is situated. Therefore certain interest groups — not only religious organisations but even political parties and others in positions of power — have a vested interest in promoting forms of religious expression favourable to their own positions. In some cases this can involve the legitimisation of 'official' interpretations of sacred texts and of personal spiritual experiences.

Therefore not only academic and other researchers of religion but also politicians have paid particular attention to this 'third' level of 'religious' behaviour. Partly this is because it was almost impossible to investigate other levels during the Communist period, when a very indirect measure of religiosity was provided by occasional references to religious behaviour as manifested in a public context.[13] However, this seems to be merely the 'surface' level of 'religious' or 'spiritual' expression. The Communist party probably made a serious mistake in trying to crush this surface manifestation of 'religion' without understanding the multiple roots of spiritual awareness and expression inside people. Therefore those persecuting 'underground' religious groups often found themselves in the position of gardeners who repeatedly cut off the heads of dandelions without pulling out their deep roots.

Researching 'spiritual awareness'

Understanding the deeper roots of spiritual awareness demands a more sensitive approach. Trust is required, otherwise people will retreat like hermit crabs into their protective 'shells'. To some extent the 'outward' forms of religiosity — ritual practices and official beliefs — have served as shells which can be presented to the outside world. Inside the shells are sensitive beings with deep feelings and emotions which are only revealed if they feel that the surroundings are 'safe'. Therefore the researcher needs to approach this topic with sensitivity and tact. Most of all, there needs to be the establishment of trust.

Previously I mentioned that Irina probably confided in me about her experience of the angel partly because she felt that I would not laugh at her or dismiss her experience as fantasy. We had already known each other for about six years before Irina had her experience of seeing an angel, so there had been time for trust to develop. In more formal interview settings there is less time to develop that kind of trust. Nevertheless, my collecting accounts of spiritual experiences was largely prompted by my having already conducted such a study among a random sample of nurses in Britain.[14] For such interviews it was important not only to assure people that their accounts would remain confidential but also to allow them to feel more at ease by starting the interview with questions of a more straightforward and non-threatening nature. By the time we discussed issues of personal spiritual experience, a certain measure of confidence had been built up. This was important, because at times the accounts related to such personal and emotional issues that sometimes people began to weep as they related their experiences.[15]

Confidence in talking about such experiences can also come from the simple realisation that they are not uncommon and that it is acceptable to speak of such matters. One of several instances of this occurred in September 1998 when I presented a paper on dreams and spiritual experiences at an academic conference in Kirov: afterwards two Russian women separately volunteered to tell me of their own experiences.

Any study of spiritual experience is at present more likely to be qualitative than quantitative largely because each experience is unique in itself. It is possible to impose general categories on the data — such as 'premonitions', 'visions', and so on — and then to attempt to draw general conclusions, but there is also a place for letting the accounts speak for themselves. This I have attempted to do. In describing religiosity in an area as vast as the former USSR and so ethnically and religiously diverse, some common themes can be discerned at the level of religious experience but there are of course major differences too. Perhaps on account of the influence of Communism, relatively few 20th century researchers have paid much attention to spiritual experience in the USSR — although it was an aspect of some 19th-century accounts of Siberian religion.[16] More organised, 'official' kinds of religiosity are easier to study because they are more obvious and measurable, and I shall also examine these in the later chapters of this book. My approach to the topic, however, differs from that of many who have had to write within the constraints of Marxist or certain other ideologies in so far as I seek to avoid a reductionist approach — that is, an attempt to explain away the phenomena as being 'nothing but' imagination, a psychological aberration or a sociological phenomenon. Any such interpretation or conclusion has to be based on hard evidence. My book attempts to provide at least a general overview of the kinds of issues on which further research is needed.

One legacy of atheism in Russia has been a reluctance to speak about 'religious' matters. For instance, in 1991 a lady belonging to the Komi ethnic group of northern Russia told me how I was the first person with whom she could speak about religion in a natural way. Others had either laughed at her or else had made her the subject of excessive curiosity. She had therefore retreated into a protective shell whereby she no longer spoke about her religious beliefs to 'outsiders'. Therefore in my interviews with Mongolians about their spiritual experiences I assured them that all information would be confidential and that I would omit their names from any report on my findings. In other circumstances, such as the experiences of Irina and Akhtyam, it is possible to use their names but in certain cases cited in this book I have taken the liberty of using pseudonyms rather than real names in order to maintain confidentiality.

People have to be treated as people in their own right — not as if they are museum specimens. Trust has to develop, but in relating issues of spiritual experience I suspect that it is probably harder for people to trust researchers who are known to be atheists or agnostics. Those who acknowledge a certain religious sympathy or practice of their own are in some cases more likely to gain the trust of those whom they are interviewing, especially when it relates to 'deeper', more personal issues of spiritual experience. This was certainly the case when I was conducting a follow-up study of a Christian healing conference.[17] At the beginning of the interviews some people were a little suspicious of me and of my motivations in conducting the research, but they were put at ease not only when they knew that my research had the approval of Christians whom they respected but also when I assured them that I have a religious faith of my own.

Anthropologists writing on this issue have sometimes quoted Schmidt, who expressed the view that: "If religion is essentially of the inner life, it follows that it can be truly grasped only from within. But beyond a doubt, this can be better done by one in whose inward consciousness an expression of religion plays a part. There is but too much danger that the other [the non-believer] will talk of religion as a blind man might of colours, or one totally devoid of ear, of a beautiful musical composition."[18]

Therefore a religious faith in the researcher can actually be an advantage rather than a disadvantage, as I have argued elsewhere.[19] All researchers have certain presuppositions and possibly unconscious assumptions about the nature of the world. Often they do not have well-defined positions about their own attitudes to phenomena such as telepathy, precognitive dreams, alternative medicine, altered states of consciousness and so on.[20] Therefore, when writing about such issues, it is not clear how far their own beliefs or assumptions are actually affecting their perceptions and their research. It is much easier to recognise religious influences upon researchers having a more clearly defined set of beliefs.

Anthropologists strive for objectivity, but can rarely, if ever, break free from the fetters of subjectivity. Increasingly they are recognising that the anthropologist's own background, theoretical bias, temperament and experiences of fieldwork can channel his or her thinking along certain lines and affect the conclusions reached.[21] However, instead of abandoning anthropology because it has a subjective component, anthropologists try to recognise the limits of these influences and to disentangle them. They also recognise that such an aim might never be achieved completely, but this does not deter them from seeking after objectivity.

The dilemma is well summarised by Martin Southwold, who writes: "We do think of ourselves as scientists, objectively observing and analysing the data, and we are right to do so, lest we lapse from standards of honesty and objectivity, rigour and open-mindedness. But the ideal of Science . . . is not to be embraced too literally and exclusively. We seek to be scientists, but we are also men among men, both in conducting research and in what we make of it. We study by participant observation . . . and our principal instrument of enquiry is ourselves, as human persons relating to others. In consequence, what we produce is bounded by our personal limitations; it had better be enriched by our personal assets, which extend far beyond what gifts we may have as scientists in the ordinary sense."[22]

In the social sciences it is increasingly being recognised that the very presence of an investigator, and the very fact that questions are being asked, in themselves can affect the kinds of results obtained. One might compare the problem with that encountered in sub-atomic physics, whereby one can measure either the position or the momentum of a particle (if indeed it is a 'particle'!) but not both at the same time.[23] In fact, the very process of investigation determines to some extent the information which can be obtained. In the same way, in dealing with human beings there is a similar kind of 'uncertainty principle' about the extent to which the very presence of the researcher affects the behaviour of those being described or investigated.

Previous generations of social scientists had either tended to ignore this issue or tried to conceal it by their style of writing. However, it can be argued that it is better to be open about the question and thereby to facilitate one's research by turning these apparent handicaps into assets. This process is again well described by Southwold, who writes: ". . . I have freely resorted to value judgements of people and their conduct, have attached weight to my own subjective impressions and feelings, and have drawn upon my own religious experiences... all of which social scientists as a rule sedulously avoid. So long as I tried to exclude my personal feelings and assessments from my intellectual analysis I was divided against myself and unable to proceed: this book emerged as an integration of what I had striven to keep apart. An anthropologist is a man — and a woman — and if he strives to be less than a man he defeats his anthropology. This, at least, is surely true when the topic is religion: was it not always absurd to expect to understand religion by excluding value judgements, emotions, and personal experience, which are its essence? I am indeed a flawed instrument; but so long as I strove to be a narrow social scientist I was maimed as well."[24]

It can therefore be an advantage to hold a religious faith when investigating religion because, firstly, it helps to demarcate the areas of one's own possible bias, and, secondly, it affords a deeper insight into the religious experience of others, without which one might be like a blind man trying to describe colours.

Although these kinds of issues are now more generally understood by specialists in the sociology of religion, it has been necessary to discuss them for the sake of those readers whose interests lie in other areas. At times I shall refer to my own experiences, but for the most part I shall attempt to portray as accurately as possible a relatively unexplored realm of religiosity in the post-atheistic cultures of Russia and Central Asia.

A Chukchi family in Lorino, by the Bering Straits

A Chukchi reindeer herder (back left) and members of his family at their home in Lorino

1
Peoples and cultures

The opiate of the people

"Is it morning now, or is it evening?", asked a Chukchi grandmother who came to the door of her home, spotted me outside in the street and called this question out to me. When I told her it was evening, she replied: "So why have I been asleep then?" She had fallen asleep while drinking vodka, and her husband was still drunk inside the house.

Their daughter came to the door and was surprised to find her mother talking with an Englishman. We were in the village of Lorino, next to the Bering Straits — a State farm subsisting largely through reindeer breeding. To get there had involved a long flight from Moscow to Anadyr, where I had been stuck at the airport for three days awaiting the flight to Lavrentiya. Each day it was postponed on account of the Lavrentiya landing strip being flooded by the thawing snow. I had been unable to cross the frozen bay back to Anadyr because the ice was melting and it was no longer safe for buses to drive across. Local people advised me that the helicopter service was too often interrupted by fog for me to be sure of getting back to the airport before my flight, which each day was postponed by just one day until finally it flew on the third day. Many local people were having to remain at the airport too, either staying at the hotel upstairs or else sleeping where they could on any available seats. I ended up talking with quite a number of them and conducting informal interviews about their lives.

Meanwhile, I had learned that the helicopter flights from Lavrentiya to Uelen and other Chukchi settlements were also very unreliable on account of the frequent fog, so one of my fellow passengers suggested I go instead to the village of Lorino which was accessible by a four-wheel drive 'bus' with huge tractor-like wheels. The route across the tundra involved several long diversions where melting rivers had swept away the road or bridges. Camouflaged by the snow covering the tundra all around us, a snowy owl had been devouring its prey but took off as we approached. Marmots could also be seen scurrying into their burrows. The snow covering the land merged into the ice covering the sea, with the almost universal whiteness broken only by occasional rocks and by the dirt track itself, until we arrived at the equally featureless grey concrete buildings forming the central square of the Lorino state farm.

My arrival on a Saturday meant that the local shop was being allowed to

After Atheism

*The central square of Lorino state farm, near the Bering Straits.
Opposite: Preparing a dog sled team for hunting or fishing*

sell vodka, whereas it was forbidden to do so on weekdays so that people's work would not be affected by drink. That weekend I witnessed several cases of drunkenness, and I helped to carry home one man who had collapsed in the street. On the Saturday night one local man was murdered by another Chukchi man while both of them were drunk. The following morning the victim's body was discovered on the nearby beach. Although such incidents are not necessarily too common, for a village with a population of only 1,600 people the alcohol-related mortality rate is comparatively high.[1] It is not surprising that the average life expectancy of some of these Arctic peoples is noticeably less than the average for Russia as a whole.[2]

Living with one Chukchi woman was a Russian man who was hoping to escape to Alaska one winter when sometimes the ice freezes all the way across the Bering Straits. Meanwhile he made a living by buying vodka in the town of Lavrentiya for 600 roubles and selling it in Lorino for 3,000 roubles — the prices prevailing at the time of my visit in May 1993. Local schoolteachers and the local doctor told me of families in the village where the parents are out of work but spend on vodka the state allowances intended as child benefit payments for their children. Their children end up malnourished and deprived, so it has been necessary for the local authorities to take some of these children into care at the local boarding school — which was originally set up to accommodate children of families living out on the tundra rather than those belonging to families resident in the village. There is also an ordinary school (not boarding) for children resident in Lorino, and one of its

Peoples & cultures

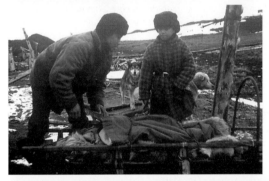

Clockwise from top right: A Chukchi man (somewhat intoxicated); Chukchi men in Lorino; Chukchi schoolteachers in Lorino; a Chukchi mother and child

teachers has organised a campaign to promote teetotalism because she sees the damaging effects of alcohol among many Chukchi families.

Similar effects of alcohol are reported a few thousand miles to the west, among the Khanty and Mansi people of north-west Siberia. There a local doctor told me of cases in which Mansi mothers have rolled over while in a drunken stupor and have suffocated their infants lying asleep beside them. Although this is not a danger in homes where the babies still sleep in traditional-style hanging cradles, there are several other causes of infant mortality which have meant that, despite many families having several children, the overall populations of some of these people groups have remained relatively static over several decades.

The same doctor said that many of these northern peoples by heredity lack an enzyme which breaks down alcohol in the blood stream. If they drink alcohol, they easily lose control of their own behaviour. I have witnessed instances of this on several occasions during my visits to Siberia. For example,

*Mansi infants in traditional-style cradles.
Bottom: A Mansi family in Lombovozh*

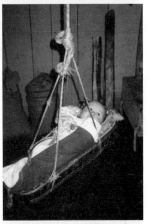

in 1995 I was in the small town of Berezovo in the Khanty-Mansi region when suddenly I saw a woman running in my direction pursued by a man clutching a bottle of vodka. The look of fear on the woman's face made verbal explanations unnecessary. I placed myself in the way of the man, who was obviously drunk. He diverted his threatening behaviour towards me while the woman made good her escape.

About a day's journey by river boat from Berezovo is the Mansi village of Sartinya, which I first visited in the summer of 1991. On our arrival we discovered that all the Mansi were drunk. They had been drinking in remembrance of a 27-year-old man who had died a year previously after falling off a bridge into a river: he too had been drunk at the time.

The roots of 'alcoholism' among these people are complex. To some extent it might be related to a lack of ethnic pride and national consciousness among people like the Mansi. However, there are also complex personal factors involved, as illustrated by a Chukchi young man

*A Mansi mother and her child.
Below: Mansi children*

named Igor who had become an alcoholic but wanted to be free from his addiction. A contributing factor in his turning to drink seemed to have been his history of having had two broken marriages. Another influence became apparent one morning when he mentioned that he had many disturbing memories from his experiences as a soldier in Afghanistan. Igor was reluctant to talk about these, but it seemed that his drinking began partly as an attempt to escape from the regrets of the past.

Atheism could do nothing to alleviate the gnawing doubts and questions which come from a troubled conscience. Inside oneself there remains a sense of moral right and wrong which seems to demand the reassurance which comes from a sense of having been forgiven. Essentially these are religious issues.

However, the propagation of atheism meant that many people had to live with their troubled consciences. Instead of religion being the 'opiate of the people' (according to Karl Marx), vodka and other drugs became the opiates of the atheists. It was through the Russians that vodka was introduced not only to the peoples of the north but also to those of Central Asia. To a large extent many other aspects of the Russian culture have also been adopted by these various peoples, but they have also retained their own traditions, values and beliefs. Some aspects of their cultures will be introduced in this chapter, considering in turn the shamanistic or so-called 'pagan' peoples of Siberia, the Muslim peoples of Central Asia and the Buddhist peoples. Later chapters will look in more detail at different dimensions of religiosity among these peoples.

Peoples of Siberia

The indigenous peoples of Siberia are the north Asian counterpart to native Americans: in both cases, the original inhabitants of the land were conquered and marginalised by the Europeans who have now become numerically dominant in these territories. It is estimated that prior to the Russian conquest of Siberia there were about 120 different 'language communities' but the 1989 Soviet census recognised only 35 distinct indigenous languages.[3] Some languages have died out but others are now regarded as 'dialects' even though they are unintelligible to each other.

These peoples have been affected in a variety of ways by the social and political changes taking place in the former Soviet Union, but the impact of

Komi Zyrians, who live on the north-eastern flanks of the Urals

Constructing a 'dug-out' canoe in the Mansi village of Kimkyasui

An Eskimo woman in Anadyr

the changes has not been uniform. My own fieldwork among such peoples is of course, like all other researchers, bounded by limitations of accessibility and time available but includes material from both tundra and forest peoples, and from both Western and Eastern (or Far Eastern) parts of this vast territory. These areas are as follows:

a) Parts of the Khanty-Mansi region of north-west Siberia, mainly some Mansi villages of the Berezovo district but also including visits to other Khanty and Mansi settlements elsewhere in the region. Some of these villages also include people belonging to the Komi and Nenet ethnic groups.

b) The state farm of Lorino by the Bering Straits, where the local people are almost entirely Chukchi, plus a few Eskimos and a few Europeans. While in Chukotka I also did some research in the town of Anadyr.

c) The settlement of Andryushkino in the tundra region of the Sakha Republic (formerly known as Yakutia). Andryushkino is an ethnically mixed settlement containing Evens, Sakha (Yakuts) and Yukagirs, in addition to a few Russians and individuals of other nationalities. Visits to Yakutsk and the Neryungri area (in southern Siberia) enabled me also to meet with those belonging to the Sakha and Evenki nationalities.

Most of the indigenous people of Siberia were forced to join collective or state farms during the 1930s. Many of them still officially belong to State farms but in practice they are unemployed. For instance, in the village of Var-yogan and its surrounding hamlets (in the Nizhnyevartovsk district of the Khanty-Mansi region), 228 people out of 532 people are officially 'employed'. In practice, however, many of them have little or no work to do. They had been employed as fishermen or hunters, but now their traditional hunting territories have been polluted by spillages from oil pipelines. All of the forty fully unemployed people are either Khanty or Nenets.

Prior to my first visit in 1991, the Mansi village of Lombovozh in the Sosva area had been on a secret list of places which should not be shown to foreigners. Even some Russians or Ukrainians have been shocked at the dilapidated state of some homes and by the fact that the village had electricity for only about two hours a day: until 1989 there had been no electricity at all.

Four days before my first visit there in 1991 a two-month-old child had died, largely on account of the polluted drinking water — especially in the

An Eskimo family (left) and a Chukchi woman (right), standing near to the shore by the Bering Straits

An Evenki man in the village of Iengra, near Neryungri in Southern Siberia, eating a meal: soup with reindeer meat, reindeer liver, bread and tea

summer there had been many cases of dysentery and sometimes a form of typhoid because of contamination of the local stream by human sewage and animal excrement. I was told that the local people had unsuccessfully tried to dig a well by hand but the sides had collapsed before they found any water. Their requests for technical help from the district administration had fallen on deaf ears until I found a Western charity willing to help supply portable drilling equipment. The Russian-dominated district administration then agreed to supply locally available materials (e.g. cement, gravel, etc.) but did not supply washed gravel of the specified grade (from 12 to 16 mm in diameter) so its effectiveness as a filter was reduced. Nevertheless, subterranean water is free from the harmful bacteria which had caused so much illness in the village. One pump was installed but plans for three more were frustrated by the attitude of a new head of the district administration who felt it was shameful to receive assistance from abroad. Local people are still waiting for him to fulfil his promise to solve the remaining problems himself.

The shift from the Communist-style centralised economy to a market economy seems to have affected to differing extents the native villages in the three areas I have visited. My impression is that there has been little change, and perhaps stagnation, in these villages in the Khanty-Mansi area. The collapse of former state subsidies and of state enterprises added to unemployment, so that cash incomes had declined in real terms. More people now are subsisting simply by local hunting, fishing and gathering. The same

Peoples & cultures

Top: Lombovozh as seen from a helicopter. Right & below: Houses in Lombovozh

A Mansi woman in Lombovozh fetching water from the river

also appeared to be the case in the Chukchi state farm settlement of Lorino.

A contrast is provided by the village of Andryushkino, where the former State farm has broken down, largely along ethnic lines, into three economic fraternities. Each of these acts as a kind of smaller 'collective farm', based on reindeer herding and fishing, but is dependent upon the former village-level structures for the supply of amenities such as electricity and water. This change was apparently related closely to the Yukagirs' desire for some kind of ethnic identity and autonomy in order to help preserve their culture and language from possible extinction. At one time the Yukagir territories stretched from the Lena river to the Anadyr river, but according to the 1989 Soviet census, there are now only 1,112 Yukagirs. Even this small population still contains two distinct linguistic groups, speaking what amount to two different languages rather than merely differences of dialect. A southerly group called the Oduls, numbering about 600 Yukagirs, lives in the forested, taiga region around the Yasachnaya and Rassokha rivers, which flow into the Kolyma, and are concentrated particularly in the settlement of Nelemnoe. Further north, in the tundra, live approximately 500 Yukagirs, called the Vaduls. Andryushkino contains almost half the Vaduls and is their main settlement. Most of the remainder are scattered throughout other settlements of the Siberian north, mainly on account of intermarriage with non-Yukagirs.

Until 1957 the Vadul Yukagirs had lived in a settlement named Tustakh-Sen, where they were engaged in reindeer herding, hunting and fishing. Then

Peoples & cultures

Above: The Arctic village of Andryushkino (in November) seen from a helicopter
Below: Even and Sakha (Yakut) people at the Andryushkino village community hall

Khrushchev's policies of amalgamating smaller villages or collective farms into larger communities meant that the settlement of Tustakh-Sen was destroyed. The Yukagirs had to move 75km westwards to the predominantly Even settlement of Andryushkino where they were no longer treated as a distinct ethnic group and began to lose their own cultural identity, language and traditions.

The recent political reforms in Russia gave the Yukagirs an opportunity to try to revive their own culture. At a meeting in 1991, the Yukagirs of Andryushkino decided to form their own economic unit within the village. This became a reality in March 1992, when three reindeer herding brigades, consisting mainly of Yukagirs, separated themselves from the former state farm and formed themselves into a separate trading organisation. Their new economic unit is called by the Yukagir name *Chaila*, meaning 'Dawn'. In

Yukagir women (wearing hats of Arctic fox fur) inside the 'Chaila' office

essence it is a kind of 'collective farm' within the state farm. The three brigades within Chaila owned about 5,000 reindeer in 1991, herded on the adjacent territories to the east of Andryushkino. Their three herds were still referred to as Herds Seven, Eight and Nine, according to the old Communist state farm system of numbering them. Tustakh-Sen lies within the grazing territory of Herd Nine. (Beyond this herd there was also an area allocated for Herd Ten, but this was not in use on account of previous overgrazing.)

After the young reindeer are born in the spring, the number of reindeer managed by Chaila grew to between 7,000 and 8,000 during the brief Arctic summer.

Reindeer on the tundra

Before winter set in, the male and female reindeer were separated and the herd culled by the slaughter of many of the male reindeer, bringing the total herd size back down to about 5,000. Some of the reindeer meat is stored for the herders themselves to eat during the long winter and the remainder is sold by Chaila, which uses the proceeds to pay its employees. About 20 per cent of the herds managed by the Chaila brigades are privately owned by individual employees of Chaila, whereas proceeds from the remaining 80 per cent are administered by Chaila for the benefit of the whole group. In addition to reindeer herding, Chaila is also engaged in fishing and some hunting.

For all these activities, however, the cost of transportation reduces profits considerably. During the winter the meat or fish often has to be transported by air from Andryushkino to the town of Cherskiy, and from there to Yakutsk and perhaps on to other cities. With rapidly increasing air fares, reindeer venison is becoming more and more a luxury meat which fewer Russians are able to afford. By 1999 their herds had halved because the herders had to kill more of their reindeer simply to meet short-term needs, at the expense of the longer-term viability of the herds.

Chaila calls itself a 'community' (*obshchina*). Most of its 87 employees are Yukagirs but there are some who are non-Yukagirs married to a Yukagir spouse. Similarly, there are a few Yukagirs who work for one of the two other communities in Andryushkino which have been set up along similar lines to Chaila, the members of which are mostly Evens. My impression, based on a handful of cases, is that those couples with inter-ethnic marriages tend to work for an association which is predominantly composed of members of the husband's ethnic group.

Some Yukagirs see the recent formation of Chaila as the first step towards re-establishing themselves as a separate group not only economically but also territorially. They had plans to rebuild Tustakh-Sen because they felt that they needed a separate settlement in order to prevent the loss of their own language and culture, but such ideas have met with less enthusiasm among other Yukagirs.

The teaching of indigenous languages is a key issue in the preservation of these northern cultures. Few of the 110 Yukagir children speak the Yukagir language. For the Yukagir children there are now some Yukagir classes at the local school (which has two buildings, one for primary and the other for secondary school-age pupils), but the majority of the lessons are conducted in Russian. Similar problems in the provision of suitable education for Mansi children were expressed to me also by the principal of the school in the village of Sosva, who said that their educational achievements are lower than the average for the country as a whole. His estimate was that less than a third, and sometimes perhaps only about a quarter, of the children can cope reasonably with the standard national curriculum. Possible reasons for this, in his opinion, were: 1) A unified national curriculum and standardised textbooks throughout the country do not take into account local ethnic particularities.

2) The textbooks for secondary schools are prepared in Russian, rather than local languages. Although there have been some textbooks published with glossaries, stress marks and explanatory notes, it has been impossible to obtain such textbooks for their village school. 3) The school is poorly equipped and they cannot provide equipment apart from blackboards and textbooks — if finances were available they would like to buy tape recorders, scientific laboratory equipment, computers and an overhead projector. 4) There seems to be a poor understanding of the psychology of the native children. This has never been studied by a professional psychologist and the teachers (many of whom are Russians) are unaware of distinctive psychological characteristics of the Mansi children. Therefore children who initially make good progress in primary school might later make slower progress because of psychological problems of which the teachers are unaware. The principal stressed that this is his personal viewpoint, but it highlights the cross-cultural problems involved in assessments of educational standards.

Yukagir children studying the Yukagir language at the school in Andryushkino

Most of the children at the school are Mansi but their native language is taught only during the first three years of schooling. When the school was first founded in the 1930s, the Russian teachers who came learned Mansi and taught in Mansi, but this process was interrupted in the 1950s. Only in the last decade has it been resumed: in 1988 Mansi began to be taught as a separate subject in classes five to eleven, and since 1991 aspects of the indigenous cultures have also been taught in classes nine to eleven. At the primary school all the children, irrespective of nationality, have Mansi classes as a separate subject, but most of the teaching is in Russian. Since 1988 kindergartens have also introduced the use of native languages. However, many of the Mansi children do not know the Mansi language: all fifty of the children in classes five to eight are Mansi but only twenty-one of them have a comfortable knowledge of the language, and the remainder know very little Mansi.

The level of literacy and of general education among the parents is relatively low. Most of the parents had attended four or five years of primary school, or at most eight or nine years. The school's principal says that parents do not interfere in the running of the school, but neither do they seem to help or encourage their children very much at home: the parents tend to leave to the school the entire burden of educating their children. It is not clear to what

Kindergarten children in Lombovozh

extent this might reflect a feeling that the educational system is one imposed on them by the Russians.

A few children leave school between classes six and nine (when they are aged eleven to fourteen) to live with their parents by fishing and hunting. Some finish the eighth class but only a small number remain at school until the eleventh year, when they

Children from the Selkup people of Western Siberia, whose language, like Mansi, belongs to the Finno-Ugric group

are sixteen years old. After school, some might go on to attend lower-grade colleges of education or schools teaching handicraft skills, but very few — described as "unique cases" — go on to universities or other establishments of higher education.

Similar kinds of situations were described to me by teachers in Lorino and Andryushkino, except that the unified national curriculum had started to break down in order to give more scope for teaching traditional occupations like fishing, hunting, handicrafts and skin or fur processing. In Lorino these kinds of traditional skills were being taught and there were classes in both Chukchi and Eskimo. At the college of education in Anadyr there are teachers of Chukchi, Even and Eskimo, whose courses are taken by Russians as well as by native students.

In Andryushkino, children aged between seven and ten who belong to the native people groups have six hours per week of tuition in their own languages; these classes include folktales and traditional songs. Between the ages of eleven and sixteen it is increased to ten hours a week. However, Dora Nikolaevna, the Yukagir teacher at the secondary school, admits that the pupils' knowledge of Yukagir is limited to what is in their textbook and that they cannot speak or read it in other contexts. Only two of the Yukagir children are from homes where both parents are Yukagirs. Dora Nikolaevna is herself Yukagir and is married to a Yukagir man, but they speak Russian between themselves at home because Dora's husband was brought up elsewhere, where he heard but did not speak Yukagir.

A new subject at the school is called 'National Culture of the Peoples of the North', which is studied by all students, regardless of their ethnic backgrounds. At the beginning the pupils learn where their relatives had come from. Then they learn about traditional pursuits such as sewing, beadwork, fishing and national sports or games. The older classes are taught about hunting.

Sakha (or Yakut), Even and Yukagir are the three native languages taught in the school, corresponding to the three main ethnic groups represented in the village. At the school there are 51 Sakha, 86 Even and 59 Yukagir children, plus 22 from other ethnic groups — including three Chukchi and three Dolgan, in addition to Russians and Ukrainians. Nevertheless, despite all these attempts to maintain an awareness of native cultures, the children's general education continues to be in Russian.

In all these native languages there is a shortage of appropriate text books. Usually the few available books tend to be limited to a first-stage language textbook, a dictionary and perhaps one or two story or poetry books. A further problem is that of dialectical differences: for instance, in Andryushkino the local Evens speak the northern form of Even but the text book is in the southern form, which means that the teachers have to explain the book to their students and try to translate the text into the local form of Even.

Among all these peoples of the north, the traditional religion focused on spirits connected with nature (e.g. gods of hunting, fishing etc.), and often a form of veneration for the spirits of deceased ancestors. During the Tsarist period many of them — the Chukchi and Nenets being notable exceptions — were superficially Christianised, but often this was a veneer on top of traditional folk practices. A Nenet man married to a Khanty woman remarked to me that they sometimes had Orthodox icons in their homes but in fact used them to pray to the 'pagan' deities.[4] In the following chapters I shall discuss aspects of these religious practices in greater detail.

Peoples & cultures

'Muslim' Peoples of the former USSR

"We don't really know much about Islam", commented Mansur, a Tatar man in whose home I was staying.

"Yes, I noticed", I replied with a smile, tapping the bottle of vodka Mansur had been insisting I share with him. Mansur grinned: he knew enough about Islam to know that officially Muslims were not supposed to consume alcohol! The same is largely true of many other nominally 'Muslim' peoples further south, in Central Asia and the Caucasus region, who have likewise been influenced considerably by Russian customs. Parodying the Communist propaganda contrasting the 'bad old days' with the liberation brought about by the Communists, some people in the northern Caucasus joke that they had been "liberated" into drinking alcohol and into dancing with women.

The Soviet system introduced significant changes to the ways of life of most of these peoples. Those who traditionally had been pastoralists, leading a nomadic or semi-nomadic life with their flocks and herds, were sedenterised to a considerable degree, usually having to be accountable to the management of a State farm. Most of their livestock became the property of the State. Among the Mongols, Kazaks and Turkmen, those tending the animals might still live in traditional round felt tents while out on the steppe, but they would also have to make periodic visits to a central location and submit reports to the farm administration. A Soviet style of management was imposed perhaps more strongly among settled peoples, who were incorporated into collective or state farms (e.g. cotton farms in Uzbekistan) or else worked in the cities in factories or offices, with occupations generally similar to those of the Russians.

Throughout Central Asia and the Caucasus Russian influence has generally been strongest in the major cities but weaker in more remote villages. Nowadays there is a spectrum

Tatars outside a mosque in Yekaterinburg

of 'types' from 'traditional' to more 'Russified' (or even, one might say, 'Westernised') individuals. This spectrum includes several different but interlinked features, namely: a) knowledge and use of their mother tongues, b) adherence to Islam, c) preservation of other traditional customs.

Uighurs in Kazakstan

To some extent these features also tend to correlate roughly with age and with rural/urban differences, but to a certain degree this tendency has been a result of the greater concentration of the Slavic population in urban areas, including the cities of Central Asia, whereas the local ethnic groups tend to be more dominant in the surrounding villages.[5] Often younger people growing up in the cities and having to use Russian much more in their daily lives tend to have a more Russified or 'Western' outlook and to have lost an active knowledge of their own language and 'customs' (including religious ones).[6] Therefore a common stereotype of 'religious' people is that they are older people from the villages who can also speak their own native language.

'Religiosity', however, is in itself a multi-faceted category, and those apparently 'less religious' younger people in the cities often have an interest in astrology or other forms of divination, read magazine articles about UFOs and 'paranormal' phenomena, and might have been involved at some time with one of the many 'new' religious groups such as the followers of Krishna or the Baha'is. 'Religious' behaviour is a dynamic rather than a static category.

Even though the Kazaks and Kyrgyz adopted a variety of social customs regarded as 'Muslim', it is generally recognised that Islam did not penetrate as deeply among them as it did among the Uzbeks, Tajiks and other settled peoples of Central Asia.[7] In contemporary Kyrgyzstan there are new mosques being built and some of the younger generation of Kyrgyz are beginning to visit them, but in Bishkek a relatively high proportion of those who regularly attend the mosques are Uighurs: it is estimated that about a quarter of the 20,000 Uighurs in Kyrgyzstan attend the mosques on a fairly regular basis.[8]

Most Uighurs live in Xinjiang province of north-west China, where they number at least six million, but the 1989 Soviet census showed that 262,199 Uighurs were also living in the USSR. The majority live in eastern Kazakstan, near to the border with China, and in Kyrgyzstan. Many of the ancestors of the Uighurs now living in Kazakstan had formerly lived in the Ili valley of Xinjiang, where they were said to have been rather lax in their Islamic observances, but after moving to Kazakstan they became noted for their piety! By contrast with the Kazaks and Kyrgyz they appeared to have been more religious, but this religiosity also declined to a noticeable degree through their exposure to Soviet atheism.

In the cities of Central Asia and the Caucasus, new buildings have sometimes obliterated the sites of former mosques or Muslim seminaries. Often local residents who have moved in from elsewhere are unaware that the site had previously been used for a religious building, whereas such knowledge tends to be more commonly preserved in villages, where the people also tend to show some respect for the sites of former holy places. For generations they have conducted religious rituals at the local graveyard, for example, so their proximity to the cemetery in itself means that there is a certain 'pull' of tradition in spite of Communist attempts to suppress it. The tombs of revered Muslim teachers, most prominently those who had founded Sufi sects, also attracted pilgrims from a wide area. Some of these were in or near towns but others were in less easily accessible rural areas. Their associations with a revered Sufi leader also made them a source of spiritual power (*baraka* — loosely translated as 'blessing' or 'benevolence'). In Turkmenistan the tombs of sheikhs such as Abul Abbas Seyyari (who died in 953 or 954) or Abu Seid Meikhani (a disciple of Junaidi, who died in 1049) had for centuries been pilgrimage sites but during the Communist period, when few were allowed to make the pilgrimage to Mecca and Medina, these locally accessible sites often became substitute foci for pilgrimage. Two or three pilgrimages to the mausoleum, near Bukhara in Uzbekistan, of Baha al-Din al-Naqshbandi (1318-1389), founder of the largest and most influential of the Sufi schools, were reckoned as equivalent to making the haj to Mecca.[9]

Some of the earliest converts to Islam had been among the oasis dwellers of Central Asia, ancestors of the Tajiks and Uzbeks, whereas the relatively superficial conversion of the Kazaks and Kyrgyz did not occur until the 19th century. Meanwhile in the Caucasus region Islam had been introduced by the Arab conquests, later reinforced by Ottoman rule, but Christianity remained strong among the Armenians, Georgians and Ossetians. Further north, in the Volga-Ural region, the conversion of the Tatars and Bashkorts to Islam occurred partly between the 10th and 14th centuries, but the process continued until the 19th century among Tatars in parts of Siberia.

In 1989 the traditionally Muslim ethnic groups of the former USSR amounted to about 55 million people. As a rough generalisation it seems that the influence of Islam has been strongest further south and becomes weaker in the north among the ethnic groups who had more recently accepted Islam. Among the Siberian Tatars, aspects of pre-Islamic shamanism, including a veneration of nature spirits and of ancestors, persisted right up until the 20th century.[10] However, the final conversion of the Kazaks to Islam in the 19th century took place not from the south but from the most northerly Muslim people group, the Volga Tatars.[11]

There are also 1.4 million Bashkort people in the former USSR (of whom almost 864,000 live in Bashkortostan, in the southern Urals region), whose language and culture is very similar to that of the Volga Tatars. By the early years of the 20th century it appeared as if the Bashkorts were being absorbed into the Tatar community but from the 1920s onwards began to think of themselves as more distinct from the Tatars. At that time Stalin's 'divide and rule' policy towards the Muslim peoples of the USSR was particularly directed towards the Tatars, who had represented the greatest threat on account of their leading role in the intellectual and cultural life of the Turkic peoples. Therefore a specially appointed committee of historians, linguists and ethnologists developed (or, in Taheri's view, 'fabricated') for the Bashkorts a literary language, a national mythology and a 'history' derived from Tatar traditions but sufficiently altered so as to appear to be distinct.[12] The success of this policy is apparent today, in so far as the Bashkorts (known as Bashkirs in Russian) want to preserve their separate identity in the face of Tatar claims that they are essentially part of the Tatar nation. It is a sensitive issue within Bashkortostan itself, which has a population of 3,943,113 containing 863,808 Bashkorts but 1,120,702 Tatars.[13]

The break-up of the former Soviet Union has meant that the newly independent states of Central Asia and the Caucasus have come into prominence. Diplomatic and commercial relations are expanding considerably with these countries, partly motivated by the potential economic wealth of the oil-rich Caspian basin (Azerbaijan in particular) and resources such as natural gas in Turkmenistan or gold, uranium, gas or cotton in Uzbekistan. The Caucasus region has been brought to public attention more through the wars in Chechnya, Armenia, Azerbaijan and Georgia. By contrast, however, relatively little attention has been paid to the peoples of the Volga-Ural region such as the Tatars, even though the 6.6 million Volga Tatars number not much less than the 6.8 million Azerbaijanis of the former USSR (though there are another 7.7 million in Iran and substantial numbers elsewhere) or the 8.1 million Kazaks of the CIS.

Within the former USSR, the 6,645,588 Tatars have at least twice the population of the Kyrgyz (2,530,998), but the Kyrgyz now have their own independent country. The 2,718,297 Turkmen and 4,216,693 Tajiks of the former Soviet Union (not counting those living in Iran and Afghanistan) now also have their own independent homelands. Partly because they are now more conspicuous, and partly because access to these areas is now easier, in recent years there has been a proliferation of literature on the 'newly independent' Central Asian peoples. By contrast, the Tatars and Bashkorts tend to be relatively overlooked largely because they do not have their own state.[14]

Most regions of Central Asia and the Caucasus were incorporated into the Russian empire relatively late, during the 18th and 19th centuries, whereas

Top: Fauziya Bairamova addressing a demonstration of Tatars in Kazan wanting independence from Russia. (The slogan says, 'The democracy of Yeltsin is a TRAP for the non-Russian peoples'.) Bottom: Tatars at the same demonstration

the Tatar khanates of Kazan and Astrakhan had fallen to the armies of Ivan the Terrible in 1552 and 1556 respectively.[15] The subsequent Russian conquests of the Urals and Siberia meant that centres of Russian population have surrounded the Tatar and Bashkort homelands. Moreover, many of the peoples of the Volga-Ural region and of Siberia have now become minorities within their own titular homelands on account of Russian colonisation of these territories and intermarriage with the indigenous populations. Many of these peoples would like to have a greater independence from the Russians, but also want to avoid open conflict of the kind which has occurred in Chechnya. Therefore Tatarstan's president, Mintimer Shaimiyev, has generally preferred to use diplomatic means to secure concessions from Moscow.

There has of course been a vocal minority who would have preferred to be more independent. In the early 1990s this was expressed most forcibly by a political party named Ittifaq (Unity), which staged a series of public demonstrations in Kazan during 1991 and 1992, calling for the independence of Tatarstan from Russia. At least a thousand people attended one such demonstration which I witnessed in November 1991. Their leader, Fauziya Bairamova, was outspoken in mentioning that the Tatars never had concluded a peace treaty with Ivan the Terrible, so are technically still at war with the Russians: therefore, she claimed, the Tatars are entitled to take whatever measures might be necessary in order to reclaim their independence. Bairamova also made public comments that at least half of Russia was Tatar territory.[16] In the face of such rhetoric, it is not surprising that Bairamova was later arrested on charges of provoking public unrest.

A Chuvash singer in Cheboksary. The Chuvash are Orthodox neighbours of the Muslim Tatars

Such statements are obviously provocative, but it cannot be denied that historically Tatars exercised their authority and influence over a much larger area than that of the present-day Tatarstan. What is far less clear, however, is the extent to which their neighbours were influenced by the Tatars' Islamic religion. They did not manage to convert to Islam the ancestors of peoples such as the Khanty or Mansi, but in European Russia there was an ambiguity over the extent to which Islam had penetrated among the Chuvash and other neighbours of the Tatars. The Chuvash are descended from Finno-Ugric tribes of the Middle Volga and also from the Turkic-speaking Bolgars, who in the sixth to eighth centuries had established a state on the rivers Kama and Volga. By the tenth century the Chuvash were already identifiable as a separate ethnic group. Akiner reports that some may have become Muslims towards the end of the ninth century, "but the majority probably adopted Islam while under the sway of the Golden Horde."[17] She notes that under Russian rule there had been intensive efforts to Christianise the Chuvash and concludes that "some of the Chuvash are Orthodox Christians, a few are Sunni Muslims" but no information was available on the ratio of the one group to the other.[18] Hostler reports that the Chuvash are "chiefly nominal Orthodox Christians, some are Moslems, and others still adhere to their old pagan faith."[19] My own visits to Chuvashia and conversations with Chuvash people lead me to believe that the influence of Islam nowadays is negligible. The majority of the 1,839,228 Chuvash people regard themselves as either atheists or as Orthodox Christians, but some pre-Christian 'pagan' practices have also survived, particularly in rural areas.[20]

In addition to the Chuvash, remnants of 'pagan' religious practices have survived to some extent among their neighbours the Mari and Udmurts, who also live in the Volga-Ural region. In European Russia there are several other peoples speaking Finno-Ugric languages, including 1,191,765 Mordvinians; Akiner writes that "it is not known what proportion of the Maris, Mordvinians and Udmurts are now Muslim, but . . . it is probable that the Muslim element has remained strong among them."[21] My own research indicates that no Muslim influence is currently discernible. For the most part, these people, like the Chuvash, reckon themselves as either atheists or Christians.

Buddhist peoples

In the vicinity of Lake Baikal, just to the north of the Mongolian border, live the 421,000 Buryats of southern Siberia who constitute the most numerous Buddhist people group of the former USSR. Their language and culture is closely related to that of the Khalkha Mongols of the Mongolian republic, and, more distantly, to the forms of Mongolian spoken in China (Inner Mongolia and Manchuria). In so far as Mongolia can be counted as part of 'Central Asia', I shall also be including material from the Khalkha Mongols in chapter two of this book.

Prior to the advent of Communism, the Buryats had been the northernmost people group to recognise the authority of the Dalai Lama, who visited their region in July 1991. A little to the West of the Buryats, in the mountains which border onto north-west Mongolia, live another formerly Buddhist people group called the Tuvinians. Their population in 1989 numbered 207,000. A handful also live in Mongolia. Traditionally they were mainly a nomadic people. Some were reindeer herders; others (in different areas) herded cattle, horses or even yaks. Most of them were predominantly

After Atheism

A Kalmyk child

shamanistic, but the feudal aristocracy were also Buddhists.

Far to the West of the Buryats and Tuvinians, to the northwest of the Caspian Sea, live the Kalmyks, who number 174,000. Their language is close to Buryat and to Khalkha Mongolian. It was in order to escape from conflicts in the region which is now Xinjiang (north-west China) that the Kalmyks migrated westwards and in the early 17th century settled in their present home near the mouth of the Volga river.

In December 1943 the Kalmyks were rounded up and deported to Siberia because Stalin feared lest they should collaborate with the advancing German army.[22] It was not until 1957 that they were rehabilitated and allowed to return to their homeland, along with Chechens, Ingush, Karachay and Balkars from the northern Caucasus. The Crimean Tatars had to wait until September 1967 for an official decree admitting that the government's accusations of treason among the Crimean Tatars had been groundless.[23] Other deported peoples included also Volga Germans, Meskhetian Turks (from Georgia), Greeks from Zhdanov on the Sea of Azov, Finns from Karelia and Koreans from the Far East.

The Koreans were deported to Uzbekistan and Kazakstan in 1937 and many have remained there to the present day, although some have moved to Moscow and St Petersburg and several thousand to the western Caucasus. Some have returned to the Far East, including Sakhalin and the Vladivostok and Khabarovsk areas.

Some of the Koreans could be regarded as Buddhists but many are basically non-religious. In Central Asia younger Koreans generally speak Russian

Top: A Kazak woman (left) buys food from a Korean vendor (right) in Kazakstan. Left & bottom: Korean traders at a market in Almaty

rather than Korean.[24] In Almaty there is a Korean language theatre but its spectators tend to be mainly from the older generation. They are also the ones who read Korean newspapers, one of which is published in Almaty, the other in Sakhalin.

Today 80 per cent of the 439,000 Koreans are urban dwellers but many live in the smaller towns (rather than in villages or big cities). In the rural areas of Central Asia there still exist distinctive Korean houses with a characteristic form of heating reminiscent of ancient Roman underfloor heating; the Korean system makes use of a fire set into the floor with tubes leading away from it which heat part of the floor and lead to a chimney at the end of the house.

Even though they have lost their language and most of their traditional culture, until

recently mixed marriages were relatively rare and there was a strong feeling of ethnic identity. The clan system has been preserved fairly strongly along with traditions of mutual aid.

There are remnants of the ancestral cult, in terms of memorial rituals for deceased relatives, although these seem to be in an attenuated form as compared with rituals in Korea itself. Nevertheless, the Koreans of Central Asia keep genealogical registers and know the identity of their relatives up to seven degrees of kinship. Sometimes they practise geomancy to find a lucky place for a tomb, and there have been cases of exhumation when they think they have found a more auspicious location for the grave. The Koreans also have rites of worshipping Ursa Major and rites at the birth of a child, which are performed either by grandparents or by other members of the older generation. Religious rites among the Koreans are generally kept within the family and not much publicised.

Some Korean doctors of traditional East Asian medicine seem to act as shamans too. Owing to the suppression of both shamanism and traditional Korean medicine, they had to have a diploma in Western medicine in order to get the right to practise. As a result, they combined 'Western' (cosmopolitan) medicine with some features of shamanic rituals and exorcism to cure illness.[25]

Many of these rituals are disappearing among the younger generation, some of whom have become attracted to Christianity. Prior to the deportation in 1937, many Koreans in the Far East of Siberia had been converted to Russian Orthodoxy but many seemed to lose interest in Christianity after the deportation to Central Asia. In recent years, however, missionaries from South Korea have started churches in cities such as Almaty where the congregations contain people from other nationalities in addition to Central Asian Koreans.

An ethnic and religious panorama

Each of us has only one life, but it would require several lives of study to be able to write comprehensively about the religious attitudes, beliefs and practices of about 100 different ethnic groups spread out over about a third of the earth's land surface. However, my experience of life in many different parts of the vast region formed by Russia, Mongolia and the newly independent states of the Caucasus and Central Asia does provide me with an overview of major trends and, more importantly, first-hand knowledge of specific situations.

All these peoples have been affected in different ways by the political and economic system imposed on them by the Communist party. Obviously there are vast cultural differences between reindeer herders in the Arctic of Siberia and oilmen on the shores of the Caspian Sea. Their existing cultural heritage and their previous contacts with the Russians have further shaped their perceptions of the world. Differences remain between old and young, between rural and urban dwellers, between more educated and lesser educated people, but there are also some overall tendencies which are common to most, if not all, of these ethnic groups. To some extent these similarities might be attributable to their common experience of totalitarianism and a centralised economy, but there are also surprising similarities in some of their worldviews which are not so obvious on the surface but which have remained within a 'folk' religious consciousness. In the earlier chapters of this book I shall attempt to address some aspects of these 'folk' worldviews, beginning with an examination of experiences such as dreams and visions.

2
Dreams and visions

Dreams among Mongolians

"It was about 2am by the time I had got in from my work at the Mongolian Parliament building. I went to bed and dreamed that an important person in Mongolia had died. In my dream the lamas at Gandan monastery were praying and using rosaries. Then I woke up: it was 3am. Later that day I went into the office as usual and told my colleagues about my dream. A few minutes later my boss was informed by the Vice-President that Tsedenbal [the former Communist Party leader in Mongolia] had died that morning at 3am. He came back and looked at me curiously and asked if I'd known!"

The 33-year-old Mongolian woman who related this account to me in 1992 was apparently sensitive to dreams as a channel of information of a kind which could not be ascertained through the conventional senses. Another such dream of hers is also worth quoting:

"In 1988 I was one of six candidates applying to do postgraduate study but there was only one available place. I had to take three examinations: one on the history of the Communist party, one foreign language exam. and one on my specialist subject, which was law. I was sleeping usually for only three hours per night because I was studying for the exams. At that time I dreamed that a very famous researcher had marked my examination papers and the difference between my mark and that of another man, who had done better than me, was only 0.06 per cent. In the morning I told my parents and my husband, who said 'That's terrible', and told me not to think about my dream because maybe it wouldn't really happen. This was two days before the examination. However, the next night I had the same dream again. I told my husband I couldn't take the exam. because twice I'd had that dream. My mother told me to go to an old man — a shaman from Hovsgol — who told me not to worry, because, he claimed, if one has bad dreams they don't actually happen in real life, whereas only good dreams really come to pass. So then I took the exam. At the examination itself I was given a choice of one of six questions; I chose a very good one for me, answered it and then left. The result was that my competitor got 48 per cent. and I got 47.94 per cent. — a difference of only 0.06 per cent. So I couldn't go on to do postgraduate work.

After this my parents and husband believed me. The shaman was wrong: he'd made a mistake. He'd told me not to worry if it is a bad dream, but I know in myself that it's true when it's about something important in my life, because I usually feel it is so."

How should one interpret these dreams? Are they merely 'coincidence'? Certain individuals might seem to read into a dream far more than appears to be warranted, whereas others might not attach any kind of 'supernatural' significance to such dreams at all. To a certain extent the interpretation of such dreams involves a subjective assessment of the likelihood of mere 'coincidence' being a sufficient explanation for the juxtaposition of a particular dream with an actual occurrence of events shortly afterwards similar to those seen in the dream.

However, at some point — which can vary considerably between different individuals — the association between the precognitive dream and the actual event can seem too strong to be dismissed as merely 'coincidence'. Another level of explanation is then sought, but the choice of such explanations is limited. Apart from 'coincidence', there are basically two principal models of explanation which are invoked in such circumstances. One is the religious option, whereby precognitive dreams in most cultures around the world and throughout history have often been interpreted as revelations from God or from some other spiritual being: for instance, several such examples are recorded in the Bible.[1] However, this kind of explanation was not officially or publicly available to those living in an atheistic Communist society. How could those having such dreams then account for them? In such circumstances the only kind of explanation available was a second option which tried to account for such experiences in terms of an impersonal 'force' such as the concept of 'psi' used by some researchers into psychic phenomena. Perhaps the lack of any religious option in explaining such experiences might have been one reason for an interest in 'telepathy' and other psychic phenomena in the former USSR and some other countries of Eastern Europe.[2] It was almost as if ESP had become an acceptable mode of discourse for describing 'supernatural' experiences which in other cultural contexts might have been interpreted religiously.

Sixteen out of the twenty-six Mongolians whom I have interviewed so far have expressed attitudes towards dreams which see them as potentially significant in what might loosely be described as 'spiritual' terms. They regard dreams as potential carriers of 'spiritual' information and perhaps even as warnings of imminent future events. Those interviewed prior to the break-up of the Communist state tended to deny any belief in God — in accordance with their Marxist atheistic background — but half of them nevertheless expressed a definite belief in a 'supernatural' quality associated with certain kinds of dreams. Although one might speculate whether or not this interest in, or openness to, dreams might have had any connection with the widespread shamanistic practices and beliefs of traditional Mongolia, one could also relate it to the widespread interest in dreams throughout the majority of human societies, past and present, in which dreams have generally been regarded as channels through which mankind might receive revelations and insight from the spiritual realm.[3]

In this regard, societies which attach little or no cultural importance to dreams are actually exceptional rather than being the norm. This 'exceptional' way of thinking has influenced not only 'Western' society but also that of Russia and of other societies heavily influenced by a Marxist, materialist ideology, including Mongolia. In so far as those Mongolians whom I have interviewed to date are all well-educated people from the Mongolian intelligentsia, at present I cannot say to what extent their attitudes towards dreams are representative of the Mongolian population as a whole. Nevertheless, the following examples provide an indication of the variety of kinds of dreams reported by these Mongolians, and of attitudes towards them:

"A relative of mine, who is more than sixty years of age, told me about her brother, who had worked as a business man. He had said to her that he was going to the USSR on business. However, the following morning she told him she had had a dream about him and warned him not to go that day because something bad would happen. He didn't listen, saying it was just superstition, and he left, but for no known reason he suddenly died while he was in a hotel abroad." (A 41-year-old man.)

"Last month I had a dream in which I was having an operation on my right arm. The next day in our English lesson the teacher taught us the word 'operation'; it was a new word to me. Usually I have dreams about things which then happen. Another instance was last year — June or July 1991 — when I dreamed of the death of the husband of my wife's friend. In fact, the next month that man did die." (A 38-year-old man.)

When asked why he thinks these dreams happen, he continued: "It's because I can see a little into the future. I don't think that God or a spirit are speaking through dreams, because I don't believe very much. Rather I think it's some kind of quality or perception in myself."

"When I was fourteen years old I had a dream in which I'd fallen down and broken my right arm. In the morning [after this dream] I went to catch my horse, and sat on it without any harness, only a saddle. The horse jumped and went fast. I fell off and broke my right arm. The dream was very clear in my mind: the next day it happened. I was very shocked because it was the next day, and so very clear in my mind." (A 41-year-old man.)

The same man also said: "When I went to Australia, I had not expected to be asked to go, but I had a dream about it and within a week I heard about being asked to go to Australia."[4] (Three others who had come to England said that their visit had been foreseen in dreams, in one case by the man's wife, but such dreams could be regarded as merely the expression of subconscious aspirations.)

A man in his early thirties said: "Sometimes my dreams come true, such as once when my parents were living abroad. Maybe I was missing them, but I dreamed of sitting with my father and talking to him. Then a week later I got a letter saying they were coming back to Mongolia for a holiday very soon. Yes, I believe in dreams."

Some Mongolians regard dreams about the loss of teeth as ominous: "Before the death of my mother, I dreamed that all of my teeth came out. At that time, I just felt it was a strong warning that something bad would happen, but I didn't know how to avoid it or what I should do — such as if I needed to go to a Buddhist temple or to say something. I just told one friend about this dream. A month or so later my mother died." (A 29-year-old woman.)

"Before I left Mongolia, I had a nightmare and dreamt that two of my teeth had dropped out. After that my cousin died and also my aunt died, a month or less later. [The cousin who died was the son of another sister of my mother, not of this aunt.] There is a saying that if there is no blood when your teeth come out, someone would die. When my cousin died, of a liver disease, I wondered if there was any connection with my tooth dream. Soon afterwards my aunt died; she was very old when she died, and had a heart irregularity." (A 27-year-old woman.)

A similar connection between actual misfortunes and a previous dream about the loss of teeth is seen in the next example: "Two years ago I dreamed that my teeth would come out. I felt that there was something wrong with me. After that I had to have an operation: the dream seemed to be a sign or prediction. I don't know if seeing the dream meant I could not have avoided the operation."[5] (A 37-year-old woman.)

When asked about her interpretation of this dream, she replied: "I know it is connected with a spirit, but I don't know what kind of spirit. Usually it happens at night, because at night spirits come to my body and say something that will happen in the future. I don't know what kind of spirits: I can't distinguish; it's a kind of dream. Before I had the operation I didn't believe in the presence of God or of spirits but after I had the operation I felt it had been predicted by the dream of the teeth and so then I believed."

Instead of teeth, a dream about meat was seen as ominous by a different woman, aged thirty-three, who said: "Before I had an operation in 1988, I'd had a dream about a lot of meat — such as sausages and so on. There's a traditional idea in Mongolia that you may get ill when you dream of meat. Similarly, when you dream of water — including a river, sea or ocean — it means you will get a cough. When I dream of water I do usually get a cough, and I do get ill if I dream about meat. Dreams of fire are a very good omen, but I've never dreamed about fire. If I have a bad dream, I pray with my Buddhist rosary that it will not happen."

Sometimes the dreams seem to be communicating information about what is happening to family members elsewhere, as in the following instances:

"When I first came to Leeds, every night I cried because of dreams that my husband, father and others had died. I have terrible dreams if I don't sleep in my usual direction: I think it's to do with the earth's magnetic field. So I changed the position of my bed but the bad dreams continued. I don't know what happened. Two weeks later I asked one of the other Mongolians to help me to move the bed once again, this time to the other corner, and after changing the position of the bed my sleep has been very good. On Saturday we changed the bed's position and on Sunday I had no dream. Previously, while the bed had been in its old position, my plans had all gone wrong. At that time, when I planned my day I expected it not to work out: I didn't want to get down to my studies. Then G (another Mongolian) came and told me that the president of the Mongolian Central Bank had arrived in London and wanted to phone me, but I have no telephone number at the students' hostel in Leeds. So I went with G and we 'phoned London. I told G that I might go to London in case the president has a letter from my

A Mongolian woman on the steppe

husband, but inside myself I felt that my husband himself had actually arrived in London. G said that was not possible, because in that case my husband himself would call, not the president. But I still felt inside me that my husband had come to London. When I spoke with the president, my first question was, 'Tell me, has my husband arrived in London?' He replied, 'How do you know? Who told you?' My husband wanted it to be a surprise when we met. I went to London and asked him about my family because I'd had bad dreams about my parents having died. He reassured me that everything was all right. After that I stopped having these bad dreams about my relatives." (A 33-year-old woman.)

"I came here to England on the 16th of January. For my first ten days I had very bad dreams about my mother but I couldn't get a telephone line to Mongolia. When I finally managed to get through, I discovered that my mother had been very ill, but had recovered." (A 27-year-old woman.)

"When my husband travelled to Thailand in November last year, I dreamed he bought me gold rings and earrings of different colours, including emeralds, diamonds and sapphires. I knew these were very expensive and he did not have enough money so I knew he couldn't buy such jewellery. However, when he returned he gave me some gold rings and diamond and emerald earrings because he had been given a lot of money by the bank which had sponsored his trip." (A 31-year-old woman.)

Dreams & visions

The following dream was related by this same woman: "My mother, who was a famous actress, had been travelling in India for 45 days but we had no information about when she would return to Mongolia. We had tried to get some information from the foreign embassy but there was no news. Then I dreamed of my mother sitting in our kitchen and telling us how India is a very poor country with many beggars. I told my father about my dream, but there was still no news that day. However, after we went to bed, there was a phone call after midnight to say that my mother would be arriving from Moscow in the morning. After she came she sat in our kitchen and spoke about the poverty in India, just as in my dream."

A 37-year-old man from Inner Mongolia, northern China, reported: "My wife is very sensitive to dreams: I'd say that about 90 per cent of her dreams are fulfilled. For example, if she dreams that her parents are ill then it turns out to be true. Her father died recently and my wife regretted that she hadn't sensed it. However, just yesterday and today her father came into her dreams and she saw him doing a lot of things: for instance, her father had his back to people who had tortured him during the Cultural Revolution. Last night she had a dream that my brother had got a new house and therefore had to live outside, in a cold place as there was no room inside! (I know this is not logical.) So therefore my wife said she thought my brother had got a new house. However, she has also dreamed that my mother had died — but this has not in fact happened."[6]

Two others reported seeing somebody in a dream and shortly afterwards unexpectedly meeting or receiving a communication from that person. In such cases the significance of the dream is not apparent until the incident occurs in real life. A 26-year-old man said that the meeting with the man seen in his dream was about a week or so after the dream, by which time the details of the dream were not very clearly remembered.

Apart from precognitive dreams, certain other types of dreams are also regarded as of spiritual significance. For example, a 41-year-old woman said: "I saw my mother very regularly in dreams after her death. So I went to a famous well-educated lama in Ulaan Baator — a man who is now the ambassador to India — who told me it was bad for my mother that I had so many dreams about her. He told me the name of a book which must be read by the lamas for me. I went to the temple and paid the lamas to read this book for me. Maybe they did do so, but in any case my dreams stopped."

When asked further about the significance of the dreams for herself, she replied: "Maybe it was because I was thinking about my mother after her death. I was often reminded of her because she lived in the same block of flats as ourselves — she on the third floor and we on the eighth. As I'm the eldest daughter, we often discussed family problems together. Also I was just missing her. I don't think the dreams were actually from my mother trying to contact me — although maybe it is a kind of contact. The last dream was her asking for a cup of tea. So then I went to the lama."

This Mongolian woman said that she had been educated against believing in any spiritual world, but the dreams were nevertheless a prompt for her to visit a lama. She also admitted: "I do believe there is a world that scientists cannot explain. It is not true what the atheists say: there is some connection between dead and living people."

In my formal interviews with Mongolians, it was easier to ask concretely about precognitive dreams than about any other dreams which might be interpreted as having a spiritual significance. Those felt to have a 'spiritual' meaning are often perceived by informants as having a nature which is qualitatively different from that of 'ordinary' dreams (which might be regarded as a kind of 'sorting out' of disparate thoughts). Some dreams, however, are sufficiently noticeable that they are felt to have some kind of 'spiritual' significance. For instance, two Khalkha Mongols and one Kalmyk man mentioned dreams of theirs in which they had seen their own funerals, finding themselves in a coffin. One man also described a recurring dream in which he was out on the Mongolian steppe searching for a spring of water but unable to find it. In each case they were puzzled about the meanings of their dreams. Dreams of this nature could easily be interpreted from a religious perspective as like 'parables' symbolising the person's spiritual condition — for example, needing to 'die to self' or needing to find the spring of 'spiritual water' which gives eternal life.

Another Mongolian man elaborated in some detail some of the spiritual concepts which he associated with dreams: "I have no religious belief, but it is interesting to me that old people say they have had dreams and maybe they are going to a certain place, and some people's dreams have predicted events very well. Personally, I experience it. I observe myself and I thought it would help me to understand some of these things. Of course dreams and superstitions and paranormal phenomena come from not understanding everything. Science can explain many things but not everything. Such paranormal [things] and superstition in its own way helped to develop Science. My belief is not absolute. Even Science can't explain all phenomena, so there may be some paranormal [events]. All dreams are not [precognitive] dreams: I think very few people, not everyone, predict something in dreams. My grandmother is now over eighty, a religious person. She told me interesting things. If a person has a very clear spirit, he can see [precognitive] dreams, but if the spirit of a person is injured or taken up, he can't see precognitive dreams. So if a man wants to be clever about everything and know all beforehand, he should keep his spirit clear and not think or do bad things."

[*Prompt question:* How do you keep the spirit clear?]

"For example, there is a superstition in Western Mongolia — I am from the Urianghai — that the hat should be in a clear place, not on the floor or trodden on. If so, it would be bad. That's one way of keeping the spirit clear."

[Are there any other ways?]

"Books. Every book should be in a clear place, not on the floor or street."

Dreams & visions

Mongolians using prayer boards and a prayer wheel at Gandan monastery, Ulaan Baator

[What about morality?]

"Morality — not doing or thinking bad things about people. Like not harbouring grudges inside. At New Year (*Tsagaan Sar*) I should go to people and be reconciled as friends and everything bad between us be forgotten."

[Do you think God might speak through dreams?]

"I don't believe in God. I think God is something artificial. I don't think about who is talking with me: God is an artificial thing. If there is no partner, you can't speak with someone: from this thinking, the idea of God came. I think I communicate not with God but with natural waves. So I communicate with nature.

"Everyone takes a stone from their birthplace if they go on a long journey. The person's organism is closely connected with the place he was born, so even if he has a stone, he can still communicate with the stone and part of his own country. Every foreigner feels it. My grandmother and father also gave me stones — a symbol. Also, when I first went to university, it was my father's present. And also it's because I like my country."[7]

It is interesting that this informant feels that he 'communicates' in some manner with nature but he rejects the idea of God by saying that it is an 'artificial' construct of that very communication process. While still subscribing to an atheistic 'orthodoxy', he nevertheless attached importance to the experience of communing with some 'Other', which in this case is interpreted as nature.

Practical considerations of time, access and other resources have meant that these interviews were conducted with Mongolians studying in Britain. Therefore these accounts are from more educated and potentially upwardly mobile Mongolians whose experiences might or might not be typical of Mongolian society in general. Nevertheless, they do provide a preliminary indication of the range of experiences reported. It is an open question whether or not such experiences are more or less common among lesser educated people, or whether there are variations in the kinds of experiences reported by different social groups. Are members of the intelligentsia more or less sensitive to dreams as channels of 'supernatural' information, as compared with other sections of the Mongolian population? If they are less sensitive, then the rest of the population are even more likely to report significant dreams. This is an issue which hopefully might be addressed by future research on this topic.

Any such research will benefit from comparisons with studies of religious experience elsewhere. Studies in Britain have already indicated that reports of the occurrence of spiritual experiences at some time in one's life are more common among older than younger people.[8] Obviously the longer one lives the greater chance there is that one might have experienced a premonition, a spiritually significant dream, a feeling of déjà-vu or some other kind of event which might be interpreted in religious or spiritual terms. If, as seems likely, this generalisation is also applicable to Mongolia, then the fact that many of those studying in England are younger rather than older people means that my sample is likely to be biased more towards those who have not had any such experiences. Therefore their incidence of reports might be seen as indicating a probable 'minimum level' as compared with the level which might be reported if more older people were to be interviewed.

However, a possible counterbalancing influence is the greater freedom accorded to religious expression since the collapse of the Communist system. When I first began to interview Mongolians about such experiences in the 1980s, they usually regarded themselves as atheists but were unable to provide explanations for their own paranormal or 'spiritual' experiences within the framework of official beliefs permitted by an atheistic ideology. Those interviewed during the 1990s have been much more open to discussing religious topics and more willing to admit to having had such experiences of their own.

Dreams among Russians and other peoples of the former USSR

Russians and others in the former USSR have also mentioned significant dreams, although I was not interviewing them systematically as I was for the Mongolians. Nevertheless, several people have spoken with me about dreams of theirs which they felt had spiritual significance and had been meaningful to them.

On a visit to Yakutsk in 1993 this topic came up during a conversation with a professor to whom I had just been introduced. He then told me of a recent dream of his which he thought involved me. In the dream he had seen a group of his colleagues sitting together with an unknown person among them, sharing out some gold. He thought that the stranger might be a foreigner. I was indeed at a table with a group of his colleagues, drinking tea, when this professor happened to come in and was therefore introduced to me. What the gold symbolised is open to speculation!

This same professor, who by nationality is Sakha (Yakut), told me what the Sakha people believe about dreams involving someone who has died: "If you see a dead person in a dream, you must not accept anything he offers you, or if he beckons from across the street you mustn't follow him, or enter a house with him. The reason is that you must keep the boundaries intact between the living and the dead, and not have contact across this boundary. If you do, one of your relatives will die soon. If you do see a dead person in a dream, the next morning you should offer a sacrifice by fire to feed that person so that the dead person does not appear to you again in a dream."

Such attitudes towards not confusing boundaries between the worlds of the living and the dead is reflected in another Sakha folk belief, according to which: "A mirror should not be placed opposite another mirror or opposite a window (which has reflections like a mirror). To do so would set up a road for the other world to interfere with this one. It would be a point of contact, or of 'co-ordinates', whereby spirits from the lower world could interfere with this world and cause illness."

Therefore the Sakha people attach importance to the maintenance of clear and distinct boundaries between this world and the next. In probably all cultures there are distinctions between certain conceptual categories such as 'clean' and 'unclean', or 'sacred' and 'profane', although the nature of the classifications and the items put within them can vary considerably from one culture to another.[9] This is why a 'folk belief' such as the one cited above about the placing of mirrors can actually become an important window through which one can see some of the categories and distinctions which are relevant to people within that culture.

A Sakha (Yakut) man playing a traditional musical instrument, called a 'khomus' (a kind of Jew's harp)

Sometimes apparently puzzling comments can also yield insight into the worldview of different peoples. An example of this occurred during my visit in 1991 to a poet named Yuri Vella, whose ethnic background is Forest Nenets. Yuri put us up in a new house built in traditional Khanty style for his open-air museum in the village of Var-yogan. The following morning when we awoke, Yuri's first question was whether or not I had dreamed anything. I had slept soundly and could not remember any dream at all, but I was curious why he had asked the question. Yuri explained that we were the first people to sleep in the house since a shaman had performed a kind of 'house dedication' ritual there. If I had remembered a dream it might have been significant.

A Russian man named Sasha, living in the Arctic regions of north-east Siberia, connected dreams with his experiences of déjà-vu. He said: "I often have the experience in which I feel that something has happened before, and then I remember that I had dreamed about it previously — maybe a week before, or even perhaps a month or a few months previously. When the event happens, I remember that I had dreamed about it, but in ordinary life I usually forget my dreams shortly after waking up, apart from some particularly impressive dreams. I know I have had dreams which have then happened in real life, but I can't remember the details."

I met Sasha while we were both visiting a group of reindeer herders living in a tent on the tundra. With us was a man named Christopher who belonged to the local Yukagir people. Christopher told me of experiences similar to déjà-vu, whereby he would be thinking about something and would then see

Dreams & visions

A Sakha woman at the airport in Yakutsk

in reality exactly the scene he had been thinking about. An example occurred while we were together. Christopher said: "At 7pm, when I came in here, I had been thinking about Sasha, not only that I'd meet him, but that he'd be lying there in just the manner he actually was when I came into the tent." When asked how he interpreted this experience, Christopher said: "I don't know. Is it God? No — I'm not a believer".

An apparently non-religious Russian man whom I met in north-west Siberia told me of a dream of his in which he had been (in his words) "in the presence of the Lord". The words given to him by "the Lord" were: "You are sleeping." This man asked me what such a message might mean. All I could suggest was that it could be a metaphor for the man's own spiritual condition.[10]

A 40-year-old Russian woman in the Kirov region recounted: "What I saw in my dream all fully happened a month or two later. My three-year-old daughter and I were walking along the street on a clear sunny day when suddenly my daughter started to cry. I asked her what had happened and she said that something had got into her eye. I looked in her eye and saw that an insect had flown in, so I cleaned it out and we carried on further. That was the dream.

"Then about two months later on a clear sunny day with blue skies we were going to go swimming in the river and as we were crossing the street I had the feeling inside me that now what I had seen in the dream was just about to happen. And then suddenly my daughter burst out crying and said: 'Mummy, something's fallen into my eye!' I wiped it out and we carried on. But it was detail for detail exactly as I had seen in my dream.

"I can't explain it at all but I think it was an intimation of what was to follow because of something else that has happened already six years ago in 1993, when I dreamt that my father had died. He was living in Leningrad but wasn't officially registered as a resident. However, it was the beginning of privatisation and he wanted to buy an apartment which he could pass on to us and our children. I had a dream that he had died but I also felt that nothing would come of his ideas to obtain the apartment and pass it on to us. Some time passed and I got a letter from him. I was of course very glad, because it meant that what I'd seen in my dream hadn't occurred and that he was still alive. The letter was dated 14th April 1993. We received it about 10 or 14 days later, so towards the end of April, before the beginning of May. I had thought I wouldn't write to him but would phone him to wish him a happy May Day. He said he'd come but he never showed up. We called him again and he said he'd come but didn't. Since then we've had no news of him at all. I don't know whether he's alive or not, or whether he died because of that apartment, but again it happened that what I had seen in my dream took place."

This woman later said that she had previously been an atheist but since her experience has started to believe that there is a God.

A Russian Orthodox woman in Moscow told me of the time when she was very interested in UFOs. She then saw one in a dream but at the same time heard a voice, which she believed was the voice of God, saying: "Have nothing to do with this." Shortly afterwards she came across a magazine article by an Orthodox priest who wrote that UFOs were actually "envoys of Satan."[11]

Significant dreams also figured in the life a Russian woman named Vera from the city of Yekaterinburg, in the Urals region. When I visited her home in 1994 she was trying to observe Lent in the Orthodox manner by abstaining from meat and fish. During our conversation she mentioned a dream of hers the previous year which she had interpreted as being divine guidance.

She said: "I wanted to change my sewing machine and I saw an advert from someone who would exchange a shelving unit for a sewing machine. The night before I went to see the shelves I had a dream of various kinds of crockery and a kettle on some shelves — but the crockery was not of the kind which I had in my own flat. When I later went to look at the shelving unit, I saw arranged on the shelves exactly the same kind of plates and crockery as I had seen in my dream. This was a confirmation that it was right for me to get them, so I exchanged the sewing machine for the shelves."

Vera at that time was 23 years old. She had not been brought up as a religious person but had come into a faith during her time as a student. A dream had also been significant in this process. She said: "In my first year at university I had been interested in Krishnaism. Then in my second year I began to get interested in Christianity . . . I had to work late at night at the library so I could read the Bible when others were not around. Then . . . in 1992 I had a dream in which I knew that I was a sinner. This had the effect of making me want to reform certain aspects of my life. Around about that same time, in the summer of 1992, I had another experience when I felt that I had to go to an Orthodox church. There I felt the presence of God and I just knew that Jesus had taken the punishment for my sins."

Dreams have also featured in the conversion experiences of people from other ethnic groups of the former USSR. For instance, in 1997 I met in Kazakstan a Uighur woman named Nadia who had become a Christian. A month or two prior to her conversion she had seen in a dream a man whom

Dreams & visions

she had not met previously but who later turned out to be the pastor of a Uighur-speaking Christian fellowship.

Nadia's brother had already become a Christian and had given her a New Testament to read. However, it was in Russian and she refused to read it because she regarded Christianity as a religion of the Russians. A little later she needed to have medical treatment but did not have the money to pay for it. Nadia then prayed, asking God to help her — if he existed. Within a week an uncle of hers unexpectedly brought her more than $1,000 — enough to pay for the medical treatment.

Uighurs in Kazakstan

Nadia then began to read the New Testament but soon after she started to do so she began to have encounters at night with evil powers which were attacking her and beating her up. Sceptics might interpret these as nightmares but Nadia's relatives said it was a punishment for her reading the Christians' holy book, and told her to stop reading it. However, Nadia felt that she had to continue reading about Jesus, and sensed that the attacks were not from God but from evil powers trying to deter her from what she was doing. Finally she decided to become a Christian. She also prayed for God to protect her from the attacks, which immediately ceased.

Déjà-vu

Several explanations have been offered for the common experience of déjà-vu —the feeling that one has experienced something before but cannot say when or how it previously occurred. One theory suggests that it arises from a temporary disjunction between the two brain hemispheres, so that what is seen through the left eye and transmitted to the right brain hemisphere is slightly out of synchronicity with what is transmitted to the left brain hemisphere through the right eye. However, this does not explain a few cases of 'precognitive déjà-vu' reported by some of the nurses whom I interviewed in Leeds, in which they were able to predict what was about to happen. For example, one nurse who had not visited Ireland previously was able to describe accurately to her husband what they were about to see round the next corner.

Another possible explanation for déjà-vu is that it arises from precognitive dreams which have been forgotten but which evoke feelings of déjà-vu when the events take place which had previously been seen in a dream. In cultures which attach little significance to dreams, it is common for people to forget their dreams shortly after waking and not to give them any further thought.

About two-thirds of my Mongolian informants reported having had feelings of déjà-vu, but often their accounts were rather vague. A few of the more detailed ones are as follows:

"In April it happened to me, with our English teacher. I didn't think about it much — it just happened in a flash."

"Since coming to Leeds I've had déjà-vu two or three times, like I've seen this street before, or exactly this building before — everything is very familiar to me, but I'm one thousand per cent sure that I've not been in Leeds before!"[12]

A 20-year-old woman recounted a link between a feeling of déjà-vu and a precognitive dream. She said: "I don't often have dreams of this kind [i.e. precognitive ones], but I have had them several times. Once, when I was 13 years old, I went ice skating with some friends. On the way to the river, about five minutes from our home, where we skate on the ice, there were a lot of boys who were calling to us. I can't remember what the boys were calling. At that time I felt that this had happened before, and then, when I thought about it, I realised that I'd already seen this event happen in a dream. I was surprised about it. This happened maybe one or two months after the dream. At the time it seemed a long time previously. I can't explain this."

Several other informants connected their experiences with precognitive dreams. For instance, a 29-year-old woman remembered an experience of déjà-vu while talking with a friend on a street in Ulaan Baator. She felt it had happened previously but wasn't sure whether it might have been in a dream. Similarly, a 24-year-old man said: "I often get dreams of this kind. For example, once in Ulaan Baator I was walking in the street past a little shop and suddenly saw a small man. Then I suddenly felt that I'd seen this person before, in a dream. I didn't remember when I'd had the dream but I was sure I'd had it."

When asked about his own explanation of such dreams, this man remarked: "I believe that in dreams God shows people what will happen in the future, in order to help people."[13]

Nature mysticism

One type of religious experience is nature mysticism. However, those Mongolians whom I asked about such experiences tended to give rather vague replies and it was often difficult to decide whether or not they were referring to 'mystical' or 'natural' experiences. The following quotations illustrate this difficulty:

"In the spring, when the snow has suddenly gone and there's sunshine. Also when I suddenly notice the trees in green as the sunlight catches them — I wonder why I hadn't noticed it five minutes before, when it seemed brown. I suddenly notice that it's green, and — not colours — harmony."

Dreams & visions

Part of the Mongolian steppe

"When I'm in the countryside, with no houses or human beings around, I feel very close to nature; I feel free like a bird or a horse. Like [for example], when I'm sitting on a rock — I don't know, it's a very exciting feeling sometimes. It's a very usual feeling for every Mongolian. I like my countryside very much and to sit on mountains or in the forest, to sit and relax. I like to be in the country by myself, far away from talking with people. The feeling is very — I can't explain: it's like a feeling of part of nature, of unity with it. I feel very happy and my mood is very happy; I feel very happy with myself and my surroundings. I like nature in England too, and feel very happy.

"When I go to the countryside I feel this nature — just happy. It's a sense of nature — I go on holidays, or at weekends, to nature. It's not so romantic or sentimental as when I intend to go to nature. It's in our blood — we prefer to go to the country to feel it — to sense the nature . . .

"I feel I'm a part of Mongolia. I'm very homesick, and when I'm far from home I feel this very strongly. In different parks and in the countryside I admire the beautiful flowers and when I see this I always miss Mongolia, because it's always very close to my heart."[14]

Intuitions about a 'greater power'

Several people spoke in general and vague terms about an awareness of spiritual 'forces' or 'powers', as illustrated by a few quotes from Mongolian informants:

"I don't know what it is, but I do feel there is some power greater than individual man and nature . . . [I believe in a continuing existence after death] because it is a purely logical question: our spirits come from somewhere and must go somewhere, so there must be a life after death . . . I feel there is something beyond the material world.

"I have a feeling of something guiding: I don't say it's God — maybe Fate; it's a feeling of something supernatural . . . I don't go against what's higher than me (Fate, or whatever), but if it goes against my interests, I go against it.

"All my life is something extraordinary at times. There is a Fate — it exists in reality . . . It means I have some experiences of this sort, not big events but little ones — but they are very important for me . . . The power didn't help me in big problems but in little but important problems; [for example], I'm not used to big towns and I often go the wrong way, but it suddenly occurs that I find my way . . . [I believe in some kind of a greater being or power because] all my knowledge and intelligence says so . . . It's a continuous experience of my life, a knowledge which says that in the Universe there must be a very abstract law which is in the Universe and in our life [etc]."[15]

Apparitions or visions

Whereas dreams occur during sleep, there are some visual experiences which occur while one is awake. The term 'vision' is often used to refer to an encounter with God or an angel, whereas many accounts focus on beings believed to be 'ghosts' or apparitions. Some examples are as follows:

"When my wife was very ill, I thought, 'God, help us'," related a 38-year-old Mongolian man, but shortly afterwards he explained what kind of god he had in mind. "In the place where I was born, there is a big mountain about which old people say there is a body of a woman who was on a yellow horse: it's a god who sits on the horse. I think about this horse and it helps, and sometimes I see a person on a horse who suddenly disappears. When my wife was ill I thought about the horse and it helped: I didn't pray outwardly but inside. I just do it sometimes in difficult circumstances."

Several Mongolians recounted tales of people they knew who had encountered ghosts or been affected by other kinds of spiritual powers. Most of these are second-hand accounts rather than personal experience. For example: "When my husband's grandmother was young and had a baby, the baby often got ill. Once she was in hospital with the baby and one night a 'creature' came and wanted to take the baby away. It was a creature with a white shroud covered over it. She didn't describe it in detail but it was as if it was some creature covered with a white sheet. It was a 'being' — I don't know if it was like an animal or a human. It wanted to take away the baby. My husband's grandmother told it to go away. She said, 'Go away!' The next day the baby started to get better." (From a 32-year-old woman.)[16]

Contrary to what one might expect, very few of the spiritual experiences related to me occurred in a 'religious' setting such as a church. However, some people do sense the presence of God or receive miraculous healings, speak in tongues or have other such experiences in a church context. An Orthodox priest named Boris Razveyev told me of an occasion when he was serving the communion and had a vision of the Eucharist bread and wine becoming the body and blood of Jesus: he then remarked that this was one reason why he is Orthodox (rather than another denomination).

An Armenian Christian named Khachik Stamboltsyan reported a visionary experience which was also taken as a confirmation of the reality of religious truth. He said: "When I accepted Christ, I didn't have a gospel or New Testament because at that time it was very difficult to obtain such literature. I was already visiting church and one day we had the great festival of the Resurrection of Christ (i.e. Easter). In the evening my mother and I were at the church and I was praying on my knees when I looked up and saw above me Christ with his arms stretched out. He was in a golden, yellow light with long garments but he also had a golden belt around his chest. I looked at him and couldn't understand why the belt was there. It was a year later before I obtained a New Testament and in it I read the Revelation of John in which is written that he had a golden sash around his chest [Revelation 1:13]."

A sphere of light was described by a Turkmen woman who said: "In the middle of June 1995 my sister and I were watching television one night when I had a feeling as if someone was watching me from the window. There was interference on the television and my sister commented that something seemed to be wrong with the air-conditioner. I looked up and saw a bright yellow light outside the window, and it seemed to have eyes in it which were looking at me. Both of us saw it. When we looked out the window we saw a light without a definite shape which then descended down from the block of flats in a flowing motion and then disappeared. It was no more than half a metre in diameter. We told others about this but they wouldn't take us seriously or believe us."

In the Arctic regions of Siberia, several Yukagirs said that they had seen strange lights in the Andryushkino area which they interpreted as UFOs. Two Yukagir men testified to having seen such a light at the same time but the men were about a kilometre apart and saw it from two different directions. In 1970 a woman twice saw unexplained lights in the sky above Andryushkino. On the first occasion she was at the slaughterhouse for reindeer and a few days later she saw a light moving from side to side above the old school before it burned down. Lights of this kind have also been reported to me by Bashkorts in Ufa.

A Yukagir writer named Nikolai told me of the evening in early January 1982 when he was alone in his room and lying down for about half an hour from 7pm till about 7.30pm. The lights were off and through the skylight he could see the starry sky. Suddenly a silhouette like a head with dark blue shadows for 'eyes' appeared in the skylight and said: "Don't forget about us: write for us." The silhouette then disappeared, but as a result since then Nikolai has started to write science fiction.

After Atheism

A Mansi man named Pyotr Yegorovich recounted to me an experience in which he saw a ball of fire which he interpreted as a god. He said: "When I was on a reindeer sledge near Igrim I saw a ball of fire, flying like a star, approaching me. It had sparks flying off both sides: I understood it to be a horse with sparks flying from it. I saw it just for a short time, about 20 years ago, when I was in my fifties. The ball of fire appeared in front of me, then disappeared."

Aliya playing the piano: on top of the piano are pictures of her deceased mother

Pyotr's own interpretation of the event as follows: "I believe it was a god who passed swiftly by me and would not wait. Nowadays a god will never show himself in his entirety to a man: only the people who lived before us, our ancestors, could see him as a god on a horse, but modern men can see only the fire, not the god himself, because there's only a few years now till the end of this century and of this age. (That's why the Mansi live much worse now than they did before.)"[17]

In Kazakhstan, a 41-year-old Kazak woman who had been living with her mother told me of an experience which occurred within a few months of her mother's death. Aliya was in her living room when she saw what appeared to be a circle of light, about the size of a bracelet, which was coming towards her. It went up her arm and around the back of her head, then disappeared. Aliya interpreted this phenomenon as a sign from the other world, perhaps an indication that her deceased mother was somehow still relating to her daughter.

It should also be stressed that the kinds of people whose accounts have been quoted in this chapter are not people with psychological problems. They are normal and respectable members of society, including (among those quoted here) a medical doctor, a politician, a teacher, a high-ranking civil servant and others with a university education and professional training, in both art and science subjects. In the majority of cases they were officially atheists at the time of their experiences. Some now have a religious commitment of some kind but others are still unsure what they believe.

An Armenian young man whom I shall call Reuben described experiences of his which are borderline between 'dreams' and 'visions'. He said: "In a dream I

Dreams & visions

was climbing up Mount Ararat and at the top saw Noah's grave. There was a special stone with something written on it, and someone was with me. Another day I saw in a dream the temple of God. In front of it was a lake with very pure clear water. I swam there. Later I was in the city — it was very light and all the people around me were very glad. I heard music and song: all the people were singing it and looking at me. I still remember the one phrase of the words of the song: 'There is a needle and you can touch the sharp end of the needle; don't be afraid to touch it because a power you don't expect will help you and lead you.' That was four months ago — that is, two months before my repentance.

"[Two months later] I agreed with some friends to go to church but before that I had been kneeling at home and asking God to fill me with the Holy Spirit: I was in a state of not sleeping but also not waking, when I saw a vision. There was a big crowd and someone was preaching about the name and glory of Jesus. I was standing among them. All were laughing at the preacher and teasing him but I wasn't. This person was confused but then decided to show the cross — a big white cross. I was looking at the cross and heard a voice, saying 'Jesus is powerful': I felt a very strong power so that I could fall down. He was like a very noble person but 1,000 times stronger: I can't describe it in words. Then I heard a second sentence, saying, 'Jesus is humble.' The only question I asked was, 'How can a person be so powerful and humble at the same time?' Tears flowed from my eyes; I would call it love. Then I woke up and my pillow was full of tears — I was still crying with tears running down my face.

"When I woke up my heart was so light — a wonderful thing. At that time it was very hard to get transport [during the economic blockade of Armenia] and I didn't know how to get to church, but I asked God to give me a bus. A bus came soon: I felt my way to God was blessed. At church the preacher gave an invitation: I went to the front and I got Jesus — a wonderful thing. It was nice, as if all my problems and difficulties had been given to him, as if someone was saying, 'Don't worry: it will be all right.'

"My Christian friends had warned me that powers of evil would become active from this moment. The next day was very nice: I just expected an attack. So it came eventually: a very charming and fascinating opportunity for contact with a woman. The evil one knows the weak places in our personality, so there was this lady — the moment I was thinking this there was temptation. So I was deceived and I did it.

"Before that, I'd many times had sexual relations but this time I afterwards felt very dirty and very bad, like I wanted to jump out of the window. During sexual intercourse with this woman, I had a very strong bleeding and pains in my genitals like I'd never had before. The wounds were painful, but not from anything contagious; it was as if God was showing me it was wrong. Maybe God understood that with my own power alone I could not overcome [my weaknesses or temptations], so He put a physical barrier there. Because of the bleeding I made it very short. I had a feeling of disgust — that I was very dirty and bad. I afterwards knelt and prayed, asking Jesus to forgive me. He healed me and forgave me. I then felt wonderfully that God had forgiven me."

Such accounts indicate that spiritual experiences of one kind or another are not uncommon among many of the peoples of Russia or Central Asia. They are one of several factors which predispose people towards an interest in religious issues. Usually, however, the experience is a catalyst prompting people to find out more — such as through reading literature on spiritual topics, discussions with those whom the Russians call 'believers' or by attending a religious meeting. Experiences such as dreams and visions are catalysts prompting a choice, but in themselves they do not automatically make a person 'religious'. Often they operate in conjunction with other factors to influence people towards a religious faith, but there are many other kinds of influences which are also at work. Some of these will be addressed in the following chapters.

However, the material already presented in this chapter gives a preliminary indication of the diversity of 'spiritual', 'paranormal' or 'religious' experiences reported by people from different parts of Russia and Central Asia. My general impression is that at least half the people with whom I have discussed such issues admit to having one or more experiences of this nature — often through a significant dream. The extent to which they are willing to admit to such experiences does depend partly on the establishment of trust, as mentioned in the introduction to this book, which makes a quantitative statistical analysis more problematic. Nevertheless, the fact that sixteen out of twenty-six Mongolians regard dreams as significant in 'spiritual' terms, and about two-thirds of them reported having had an experience of déjà-vu, seems to be similar to the kinds of frequencies I have encountered in less formal interviews among representatives of other ethnic groups. Even what might appear to be a hindrance to research on these topics — namely, a refusal to relate experiences of precognitive dreams — is in itself an important ethnographic fact. This was illustrated to me in 1998 by a woman in Syktyvkar who is half Komi and half Russian. She admitted to having had precognitive dreams on some occasions, though infrequently, but said that they had always concerned something bad and so she did not want to talk about the details. This conversation took place over a meal in her home at which her husband was the only other person present (apart from myself); moreover, I had first met this woman two years previously and we had corresponded a little in the meantime so I had thought that there had been time for at least some measure of trust to develop. Her behaviour was nevertheless a further indication that precognitive dreams are not uncommon but the fact that they might relate to sensitive areas of a person's life means that there can be a tendency towards an under-reporting of their incidence. Furthermore, the very fact that this woman was reluctant to speak of her experiences seems to suggest that for some reason she was afraid of speaking about it. Certainly fear is frequently a factor which motivates some kinds of religious behaviour, as will be discussed further in chapter three.

3
Fear

Having travelled about 66km across the frozen Arctic wastes, Nikolai was glad he was approaching the tent where he would be visiting friends and spending the night. Made of reindeer skins, the tent always had a fire burning inside to keep the herders warm, although by 5am, when the fire had burned low as the herders slept, the winter cold could be felt even through the thick sleeping bags of reindeer fur and whatever other clothes were being worn.

This time Nikolai was in a hurry. Already the brief few hours of winter daylight had passed and night was falling. His wife was with him, on the sledge which he was towing behind his motorised 'skiddoo' — a kind of motorbike with caterpillar tracks. They wanted as soon as possible to reach the tent where they would be welcomed by a fellow Yukagir family and served a nourishing warm meal, mostly consisting of reindeer meat.

A few kilometres from the tent Nikolai's family passed near a tomb belonging to a Yukagir prince who had died in the early part of the 20th century. Normally they would stop to pay their respects at the graveside by laying a cigarette in front of it. This time, however, Nikolai was in a hurry and made a quick decision not to bother stopping at the grave.

They had gone only a kilometre or two further when suddenly, without warning, the engine failed on the skiddoo. Nikolai's wife was thrown forward off the sledge, her legs colliding painfully against the skiddoo. The violence of the impact was so powerful that they feared lest she might have broken her legs. Slowly and gingerly she managed to get up from the ground, but both of them were shaken by the incident.

Nikolai immediately remembered that he had neglected to put a cigarette at the grave. He retraced his route and put the offering at the grave. Here on the tundra, where there are no flowers, it might be interpreted to some extent as merely a token of respect, comparable to the laying of flowers on a grave in Europe or America, but there was another dimension to the custom too. Nikolai was convinced that his accident was a judgement upon him from the spirit world because he had shown disrespect to the spirit of that Yukagir prince.[1]

Ever since that time Nikolai has not failed to put an offering at the grave whenever he has passed that way. He also insists on others doing so too: their potential disrespect towards the spirit could have potentially disastrous consequences for the whole group of travellers.

Clockwise from top: A frozen river in the Arctic with a herd of reindeer grazing on the opposite bank; a Yukagir man outside the yaranga with his hunting rifle. ('yaranga' is a Chukchi word but in Yukagir the tent is called 'vadun nime' ['Yukagir house']); fetching buckets of ice from a lake to melt for drinking water; sledges and motorised skiddoos outside a yaranga on the tundra

Fear

In the harsh conditions of the tundra, placating or showing respect to the spirits can be regarded as a matter of life and death. While on the tundra I was told that there are certain places where it is imperative to stop and place offerings before the representations of the gods.[2] These are special sacred places, but the tents and some homes are also locations for making offerings to the spirits. On entering a reindeer tent (*yaranga*) of the kind which Nikolai was visiting, one is usually welcomed with a meal and the guest is expected to bring some vodka as a visiting present. Then the guest has to put a spoonful of the vodka onto the fire — as an offering to the fire god. Failure to comply with this custom is seen as a serious breach of etiquette. Moreover, any storm, illness or misfortune which occurs after the failure to observe the ritual may then be interpreted as judgement from the spirits.[3]

Nikolai's account of his accident shows how fear of possible reprisals can be one of the basic motivations behind the performance of certain religious rites. Among many other ethnic groups too, offerings to the spirits can have both positive and negative motivations: those presenting the offerings hope for positive benefits and help from the spirits but they can also be motivated by a fear of the spirits' anger or vengeance if the offerings are not given.

Inside the yaranga

Motivations of this kind also influence some Mansi people when they attach to trees in the vicinity of a graveyard bundles of goods wrapped up in blankets as offerings to the spirits. An Udmurt woman mentioned that fear had also been a motivation behind their traditional customs of presenting offerings of food to the spirits of the dead: although partly it was a sign to the deceased that they are not forgotten, it was also at the same time motivated by a belief that spirits of the dead would protect the family or community from illness or other misfortunes. Therefore it was important to give offerings to the dead on specific occasions in order that they would not be offended and withdraw their goodwill. Many peoples also have customs of making a donation to the spirits of a locality when passing that place: for example, a Kalmyk woman throws a coin out of the window of a vehicle when she crosses the border out of her native territory to an adjoining one. Similarly, in north-west Siberia, a Russian man throws a coin into the Sosva river whenever he passes a Mansi sacred place, as an offering to the spirit of the locality and to ensure safety in travel. He observes the custom even though he himself does not belong to the indigenous ethnic groups of that region.

Curses

"I feel that my family is under a curse," said Konstantin, and explained: "I am the last male in my clan, and my only child is a girl."

Konstantin is a 34-year-old man who belongs to the Khanty people of north-west Siberia. "There are two possible reasons for the curse," he continued. "One is that my own mother broke a Khant tradition by leaving her husband and marrying again while her first husband was still alive: probably the other relatives cursed her for this. Within a very short time after my mother did that, my two younger brothers died and I was left alone at the age of twelve and was brought up by my sister.

"I have one other male relative in my clan — an uncle who also has no male children. He probably knows more about our family and what happened to it. In fact, he even did some research and collected material, but nobody, not even his own wife, could persuade him to write or publish it. He believed it was something not to be told to outsiders.

"Another possible reason for the curse on our clan comes from the sins of our ancestors. Probably too many people, including outsiders, began to come to our family's sacred places in the forest. It might be for this reason that the sacred places all burned down within a short period of time, through forest fires."[4]

Offerings to the spirits, wrapped up in blankets and sacks, are tied to the trees near a Mansi graveyard

Fear

Konstantin works in the town of Khanty-Mansiysk, which lies at the confluence of the rivers Irtysh and Ob. Although he is one of the more educated and urbanised Khants, his outlook on life is still shaped by Khanty folk beliefs. Traditionally the Khanty lived by reindeer breeding, hunting, gathering and fishing. A few still pursue their traditional way of life in remote parts of the forest, bringing up their children without schooling and without learning Russian. Nowadays an estimated 20 per cent to 30 per cent of the Khants still continue their traditional way of life as best they can, but others now work in non-traditional occupations in the towns.

The 22,500 Khanty are closely related to the 8,500 Mansi and 34,665 Nenets peoples who are their neighbours to the north and north-west. Among the Nenets, who live mainly in the tundra regions, there is a similar idea that the extinction of a clan can be a form of divine punishment. One of their men was

Traditional Mansi 'clan idols' in the open-air museum in Khanty-Mansiysk.
Below: To the right of the 'clan idols' is a store-house for grain and other items, including household gods (figurines). (The design of the legs is to deter mice and bears from climbing up.)

known to be very sexually aggressive and would sexually force himself on a woman whenever he came across one who was alone on a dog-sledge. Although none of the local Nenets did anything to try to prevent this man's behaviour, they believe that the gods punished him instead. All his children died, then his wife died and finally the man himself died. Recently the last of his grandchildren has also died, so the clan is extinct.

Similar ideas about misfortune as a form of supernatural punishment were expressed to me by Elen Maltseva, a Khanty woman who was one of the two co-founders of the open-air museum of traditional Khanty and Mansi culture in the town of Khanty-Mansiysk. At the museum a Russian man stole some of the money which had been put as offerings in front of the head of a bear (which is sacred to the Khants and Mansi). Shortly afterwards he was involved in an accident and was hospitalised for several weeks. Elen thought that the Khanty gods had punished the man for his sacrilege.

A fear of curses can also be found among some Central Asian peoples. In Turkmenistan, there are certain people, called *'ters okan'* ('opposite interpretation') mullahs, who are attributed with the power to put a curse on somebody. For example, a wife whose husband drinks heavily might pay such a person to perform a ritual to stop her husband drinking.

Southern Nenets mixed with northern Khanty. Some Nenets regard certain misfortunes as divine judgement.

The evil eye

Some Khanty, Nenets and others wear amulets which are thought to protect the wearer from evil. When asked about the amulets, some people give the impression that they do not seriously "believe in" the amulets — but they nevertheless say that they wear them for "good luck."

A similar attitude can be found also among some Kazaks and other nominally Muslim ethnic groups of the former Soviet Union, whose amulets are often worn to ward off the effects of the 'evil eye'. Among some of the Turkic peoples there is a belief that a kind of curse can come upon a person (perhaps even unwittingly) by someone with the 'evil eye' who offers praise without first saying a protective expression such as *"tuweleme!"* (in Turkmen) or *"mashallah!"* (borrowed from the Arabic *ma sha'allah*, 'what God has willed'). It is thought that the power of the evil eye is derived from envy within the person offering the compliments, but what is not so clearly expressed is a concept that evil spiritual powers could be active in utilising the envy or the assumed curse behind the evil eye. Often there is a fear of the evil eye affecting babies, but it can also be blamed for misfortunes befalling animals or property.[5]

Despite the influence of an educational system advocating atheism, it appears as if many Turkmens, Kazaks, Uzbeks and other Central Asian peoples, and also some Tatars, do still fear the evil eye and observe folk traditions based on such fears. A Turkmen woman, for example, avoided telling anyone when she was pregnant because of a fear that she would be laughed at by "bad people" whose laughter might curse the baby, potentially causing a miscarriage. Sometimes illnesses, especially in children, are attributed to the careless comments of those who had forgotten to say a word like *mashallah* before expressing a compliment. Some Turkmen are reported to be apparently reluctant to attend parties or other social gatherings on account of a fear of the evil eye caused by what they regard as "bad vibes" from certain individuals.

Beliefs in the power of the evil eye are by no means confined to rural, elderly or less educated people. For example, a Turkmen doctor from Ashgabat was in her early thirties when she told me that "many times" she had performed a ritual to fumigate her home from the malevolent influence of the evil eye. Her medical training by no means precluded her from attributing her child's illness to the evil eye. She said:

"For many years I've had a 'friend' — or so she considers herself — whom I just can't shake off. I've told her there's no need to come round but she still comes. On the day after she comes the children are always seriously ill: whenever she visits some illness follows, whether coughs or something else, such as 'flu in the middle of summer. I don't like her and I can't accept her but I've known her for 22 years so she counts me as her friend. I feel uncomfortable when she comes, and afterwards too. She's an atheist but I feel there's an evil influence from her.

"One day another woman came to the door and I felt very uncomfortable with her. She was not a gypsy but was a like a dervish — poor, and asking for bread or money. I felt she was a charlatan, but she was demanding a lot. Now, looking back on the incident, I recognise that the woman was very unpleasant and had very evil eyes. At the very moment when I was speaking with her, my daughter — then about two years old — came through to me at the door and this woman remarked how good it was that there was a child in the home. At that time I didn't even have any money in the house and so I offered her some bread instead but she did not agree to this. I had to shut the door on her but I felt very uncomfortable about the whole incident.

A Turkmen protective amulet

"Soon after that, my daughter began to become very bad: she began to cry, had a temperature and became ill. That day I invited round a friend who knows how to deal with the evil eye. We did a test with salt which we put onto a fire: the salt flew off quickly and became black, so we knew it was caused by the evil eye - whereas if the salt had remained white then we'd have known that the illness wasn't from the evil eye. We then washed around everywhere where the woman had stood on the threshold. I wanted to wash off the evil influence. We burned a special plant and waved it around so that there was smoke everywhere, and over the heads of the children too."

This plant, called *yuzarlık*, is said to be able "to drive off evil spirits" and in some Turkmen homes is hung up next to the doorway as a protective charm. The idea that the visitor's evil influence could be washed away by water stands in contrast with a Turkmen custom whereby after the departure of a welcome visitor the hosts do not wash or sweep the house until after the guest would have arrived home.

In Kazakstan and elsewhere in Central Asia, newborn children are usually kept indoors for 40 days after birth. Some Kazaks say that this is because the infant is more susceptible to attacks from evil spirits because the child's spirit or 'energy' is weak in these early days. Blue beads are often sewn on the infant's clothing to draw away evil from glances at the baby. Besides beads, Uzbek mothers might use some cloves or a triangle of cloth containing verses from the Qur'an in order to ward off evil spirits from their children.

In Turkmenistan adults as well as children wear brooches, beads, buttons or other amulets as protective charms. Camel's hair is used for some bracelets or necklaces, while amulets made of wood or other materials are often to be found protecting the doorways of homes and certain pieces of equipment or furniture within the homes such as television sets. Sometimes they are placed on shelves near the ceiling of a room. Throughout Kazakstan and the Central Asian republics it is not uncommon to see amulets placed in motor vehicles as charms for protection from accidents.

A Turkmen bride and bridegroom outside a 'wedding palace'

Fear of the evil eye at first appears to be more overtly manifested among Muslim peoples of Central Asia but it is by no means absent among non-Muslim ethnic groups further north. To some extent it is present also among

Russians. Some people dismiss it as superstition or else see it as a component of the fashion for 'New Age' beliefs which became popularised during the 1990s: for instance, one Udmurt man commented that the evil eye has substance as a kind of spiritual or psychological influence in the form of "bad waves" which can emanate from certain people. Another opinion regards belief in the evil eye as more situational, whereby it is disregarded in most circumstances but a fear of it becomes stronger in situations of potential hazard such as pregnancy, childbirth or having an operation. Among the Russians and Udmurts with whom this topic has come up in conversation, less than half — more like a quarter to a third — dismissed the idea as nonsense whereas others were less inclined to dismiss the belief altogether even if they would not necessarily fully affirm it. Those with children of their own thought it best to be on the safe side "just in case"; therefore if their child had any contact with a stranger they would then wash the child's face with water to wipe off any influence from the evil eye. One father said he did so when his child was crying without any apparent reason and then the child stopped crying, but the father admitted he could not tell if this was due to 'natural' or 'supernatural' reasons. Unexpected illness or naughty behaviour in a child, persistent crying and cot deaths can all be attributed to the evil eye. Especially during the first month after childbirth parents try to keep the newborn baby indoors away from the sight of strangers.

One informant mentioned an "old tradition" that if a stranger has dined with the family and looked at the baby, then afterwards they should wash the four corners of the table, then use that water to wash an icon: the same water (presumably sanctified by the icon) is then used to wash the baby's face.[6] One is supposed to recite certain prayers at the same time. Sometimes salt is put on the child's fontanella to cleanse away any influence from the evil eye, as the top of the head is seen as having a direct connection to heaven. For some Russians an Orthodox christening is also regarded as a kind of charm against the evil eye.

Among the Komi people of northern Russia a fear of the evil eye also motivates some mothers to keep their newborn babies away from public view. A specialist on Komi folklore described the evil eye as something which can be involuntary but the resultant illness can result from giving praise to someone (such as remarking on a baby's good looks) — just as in the Middle East. The Komi also have a belief that deliberately intended evil of a supernatural nature can 'spoil' animals or property: examples cited to me included a cow no longer giving milk or becoming ill. However, deliberate cursing is considered to be rare nowadays among the Komi and to occur, if at all, only in isolated rural contexts.[7] One second-hand account related to me concerned a cow which became ill in a village in the south of the Komi republic. The family sought the advice of a local 'wise woman' who told them to search their home. When they did so, they discovered a walking stick which belonged to none of the family members and its presence in the home

Fear

A traditional Komi summer tent

Inside the Komi tent

could not be explained. They disposed of it and the cow recovered. As a result of this, one of the children decided that his previous atheistic worldview was an insufficient explanation for what had happened.

In general the Komi regard themselves as Christianised and claim that any remnants of paganism are now at the level of 'superstitions' or folk beliefs: one such relic of previous belief systems is the idea that one has a spirit 'double' (called the *ort*) which, if seen, is a portent of misfortune. Numerous folk beliefs at the level of 'superstitions' are commonly observed by many Russians and other peoples of the former USSR. Sometimes these are beliefs about 'lucky' or 'unlucky' events — such as the idea that a black cat crossing the road ahead of a person is supposed to be unlucky.

Other 'superstitions' influence personal actions: for instance, many people refrain from conducting a conversation or passing objects over a threshold; instead, one of the participants steps over the threshold so that both are on the same side. Some of these folk beliefs or 'superstitions' might be explained away as vestiges of belief systems which are no longer formally recognised — perhaps of the kind mentioned in the 'Secret History of the Mongols' whereby the door-frame was believed to be the dwelling place of a guardian spirit.[8]

In a museum in Syktyvkar, a pagan figurine (representing one of the traditional Komi deities) is placed adjacent to a display about St Stephen of Perm, who brought Christianity to the Komi

Nevertheless, whether or not modern 'superstitions' are relics of ancient beliefs of this kind, they nevertheless continue to exert an influence on people's behaviour in everyday life. The power of such 'superstitions' rests on a vague sense of fear — that something bad would occur if one broke the taboo.

Fears of places

"Recently my Dad told me about the cemetery in our village . . . Once (several years ago) they wanted to put up some telephone lines, but the best route for the lines was across the cemetery. For a long time the local people hesitated about whether or not to put up the lines, as it was sacrilege and not honouring to God. However, then a tractor driver agreed to do it for a large payment. He went home and got drunk so as to get courage from the alcohol, but his wife brought him some heroin along with the vodka. The man died. Now, tell me, was this divine providence or was it simply a coincidental set of circumstances?

"However, there's more to the story. The local people didn't give up the idea, but did put up this 'phone line. However, at night a strong wind knocked down nine of the telegraph poles in every direction, so that the damage caused was worse than if they had put the line around the perimeter of the graveyard. Some time later they put up the poles, but now there's a large pole standing right on the grave of my aunt's father. If even a light breeze blows, this pole hums — as if thousands of spirits are riding in on the wave. Therefore, whenever anyone is buried there, people passing by the graveyard say that the pole hums frightfully. My Dad has been there many times, as the pole was put up in his childhood.

"I don't know if I was impelled to tell you about this, but I very much wanted to relate this story to you. At home I have a lot of different holy books, and sometimes I read the Bible or the Muslim Qur'an . . ."

This extract from a letter dated 20th February 1997 was written by a Turkmen teenager whose family I had met on one of my visits to Turkmenistan. The family belong to the Turkmen intelligentsia and are professional people in Ashgabat but still have some links with this village where the father had grown up. From the tone of the letter, I suspect that this teenager is fairly typical of those who have been brought up with little or no religious commitment but is interested in the general topic of the 'supernatural'. Are there ghosts or evil spirits which can be connected with particular places? Can some people or places be affected by curses placed on them? Such questions are not merely theoretical but very practical if one is confronted by stories such as the one cited above.

A Bashkort woman commented to me that they thought their flat in Ufa was haunted. However, she had no idea what to do about it. The legacy of Communism has meant that nowadays people do not think of asking a priest or other religious specialist to perform an exorcism of a place lest they be considered superstitious or crazy. However, many Russians, including those who do not normally visit churches, will obtain holy water from a church and spiritually clean their new flat by putting holy water in the corners, starting from left to right — reported to me as being "in the direction of the sun", which sounds like there is an influence from pagan practices. At the same time the lay person is expected to command all uncleanness to depart, though these words are supposed to be said "from one's heart, believing it." To cleanse a dwelling from spiritual contamination members of other ethnic groups might call in a religious specialist or else resort to a traditional ritual of some kind.

An example of this was recounted to me by Dora, a woman belonging to the Even people of Siberia. She said: "Two and a half weeks ago my younger child was at home with the light on but saw outside a figure all in black going across to the toilet door. It was shut — but the figure walked straight through it. I went straight outside and collected snow with which I covered the area and I brushed all around the vicinity with branches of a bush with a good smell in order to get rid of the spirit. However, that same night our older son could not sleep. When I asked what was wrong, my son said he was being disturbed by the noise of someone who kept washing in the bathroom — but nobody was there. My husband was in Cherskiy at the time. However, when he came home — and I must stress that he was not drunk — he had just gone to bed and shut his eyes when he felt himself being held down and strangled. He opened his eyes and saw over him a tall, black man, but as my husband struggled for breath, suddenly the figure disappeared."[9]

A 28-year-old man from the Inner Mongolian region of China commented that he would never go to a cemetery at night and would never step on a grave. He admitted that he is afraid of cemeteries because, he says: "I feel that there might be spirits of dead people." After mentioning this, he went on to relate the following story:

"A neighbour whom I knew was on his way home one night and was drunk. Not knowing where he was, he went into a cemetery and sat down on a gravestone without at first realising what it was. When he got home, he mentioned about it to his wife and children but they didn't pay any attention to it, even though people say that one shouldn't sit on a gravestone. Then he went to sleep with his youngest son in his arms. When the son woke up the next morning, he found that his father was dead."

A further influence on his attitudes to graves comes from a story which had been circulating in his home town. He said: "In the countryside herdsmen are remote from each other. One couple had a daughter but the mother died and they buried her nearby in a damp place. Some months later, the daughter kept telling her father that her mother still haunted their home: the mother kept coming in, taking food, bowls and so on, but then disappeared. When the father woke up and found that nothing had in fact disappeared, he wouldn't believe her. This went on for some time, with the daughter repeatedly telling her father to do something about it. So eventually he did keep awake one night and did see his wife coming in and taking things away. He was frightened and realised there was a ghost. To deal with it he used a traditional method: he killed a sheep or goat and collected the blood in a bowl. The idea was to throw the blood into the face of the ghost. One night the father and daughter hid behind the doors waiting for the ghost to come in, and as soon as the ghost entered he threw the blood into the face of the ghost, who disappeared. For some days the ghost left them alone but then appeared again. The man didn't know what to do. He went to the commune centre — this was at the time of the Cultural Revolution — and reported the problem to the local police chief, a Communist who didn't believe it and told the man he was telling a lie. The man insisted there was a ghost, so the police chief went to have a look, as the father and daughter were too afraid to return to the house. The commune centre is quite far from their farm so the police chief went on a camel: the man went on a horse with his daughter in front of him. They kept on riding; it was night. When they climbed to the top of a hill they saw a light in the distance, in the man's house. The father told the policeman that the house was empty, with nobody living there, but they could see a light, so the policeman believed that there was something going on. As he approached the house, suddenly his camel stumbled in a rat hole and just at that same time the light went out. The policeman fell and broke both his legs and arms. He was taken to hospital and died a few months later, frightened to death.

"The doctors investigated what was going on. In the grave they found the body of a woman with long fingernails and long hair, as if the body was still not quite dead when she was buried. They removed her body and buried it in a high, dry place; since then nothing has happened."

This informant was in England doing postgraduate research when he related this story to me. Commenting on it, he remarked: "Rationally, I don't believe it was a ghost; I think one can explain it scientifically. Although I can explain the light by a chemical, I can't explain the vision which the father and daughter saw. Even though I've not experienced it myself, this story is so convincing that I wouldn't venture to approach a graveyard and now I'm afraid of dead people."[10]

Cultural and political roots of fear

Life in all these regions can be unpredictable, and often carries hidden dangers. Uncertainties and hazards come from many sources, including illness, the weather, wild animals, earthquakes, and other human beings. Often those facing danger of some kind seek reassurance or help from a greater power, even if they have doubts about the existence of that power. Even those claiming to be 'atheists' can find themselves in times of crisis expressing a form of words along the lines of "Oh, God!" or "God help me!" Perhaps it merely provides psychological comfort, or is said out of habit, but it nevertheless points towards the felt need for some kind of assistance from a power greater than ourselves.

Throughout Russia and Central Asia there is a widespread interest in astrology and other forms of divination. Commonly people will ask about one's star sign. To some extent it is merely a topic of conversation but there are many who subscribe to at least some kind of belief in the influence of the stars. Palmistry and other forms of divination are not uncommon too. Those

Fear

A traditional Kazak felt tent

who consult psychics and mediums are also expressing the same felt need for advice or guidance from some kind of power outside of themselves which can help them to make sense of life and provide some direction in decisions for which the long-term possible outcomes are largely unknown.

Other folk beliefs are also based on the avoidance of certain taboo actions and thereby generate new kinds of fears. For example, many Kazaks believe that one should not wear used clothing belonging to another person unless perhaps it is passed on from a family member. To do so would break a taboo and risk misfortune or illness.[11]

Fears of this kind are based to some extent on cultural classifications between 'clean' and 'unclean', between which it is felt to be important to preserve the boundaries.[12] To some extent, these boundary distinctions can become stereotyped into rules and formulae which are regarded as either 'traditions' or as 'superstitions'. For instance, in Mongolian and Kazak felt tents it was often regarded as a breach of etiquette or as taboo for boots to be put in a high place or a hat on the floor. A younger Kazak woman in Almaty told me how this perception of boundaries is carried over into city life, where her mother forbids her to put her shoes on a higher shelf in their apartment because it is "not proper."

During the Communist period all ethnic groups — including the Russians — were dominated by a political system based on coercion through fear. People feared being visited at night by the secret police and being subjected to a mock trial before being sent far away to a labour camp — or executed. Distrust became rife: people were afraid of informers who would report to the authorities any behaviour or words which were not 'politically correct'. To a large extent the sense of fear itself was used as a means for control of the population, irrespective of whether or not those fears were founded in accurate perceptions of the social environment.

It is therefore not surprising that the religious systems of these peoples are also characterised to a considerable extent by fear. However, this is by no means a simple cause and effect relationship. I do not wish to imply that the fears in the religious beliefs are any reflection of the prevalence of fear within the wider social and political environment. Clearly many of the religious beliefs and practices founded on fear predate the advent of Communism and probably also existed long before the Tsarist state. Neither can one necessarily say that the methodology employed by the servants of the political authorities in order to induce fear and subservience within the populace was a direct result of the religious beliefs based on fears. I would prefer to suggest that a basic worldview in these cultures which is strongly influenced by feelings of fear and distrust tends in itself to feed into the forms of social organisation which arise within those cultures.

In all cultures around the world there are certain natural environmental factors — whether storms, earthquakes, wild animals or other agencies — which are the focus of fears. In themselves these may or may not give rise to religious systems based on the placation of spirits believed to be behind those environmental forces. However, if such beliefs do arise, as is not surprising, for instance, in the harsh environmental conditions of northern and even central Eurasia, then over successive generations the range of fears and the types of practices which are thought necessary to placate the evil spirits are liable to become deeply engrained within those cultures.

In such a context, those who have managed to achieve political power — and might themselves fear being usurped — are not unlikely to resort to forms of despotic control based on the use of fears generated within their subjects. Fear can thereby give rise to manipulation through the use of fear.

To some extent fear can also be generated through the use or threat of force against one's religion. This has been the experience of many Tatars, who have suffered at times from attempts by the Tsarist state and the Orthodox church to convert them to Christianity through coercion. During the 17th to 19th centuries there were rather crude attempts to convert the Tatars to Russian Orthodoxy through coercion and tax inducements. At times the Russians destroyed the mosques but every time the Tatars returned to build new places of Islamic worship.[13]

In the 19th century the Russian authorities imprisoned or deported to Siberia some members of a Sufi order which advocated the restoration of the former Bulgar state.[14] These and other repressive policies of Russification pursued by the Tsars had little long-term success but led to an intensification of bitterness and antagonisms towards the Russians among many Tatars. Their Bashkort neighbours in the southern Urals also suffered from Russian oppression, including in the 18th century the handing over to slaveholders of 30,000 Bashkorts who had taken part in a rebellion. A forced Christianisation policy was undertaken among them too.[15] Again, fear was an incentive to conform and not to rebel against the status quo.

No more than ten per cent of the Tatars did become Russian Orthodox, although Muslim Tatars like to claim that the ancestors of these converts were actually pagans living in the area rather than Muslims. In any case, these Orthodox Tatars have now become almost an ethnic group in themselves, having their own distinctive customs, dress and language. They often have their own separate villages — for instance in the west of Bashkortostan, in the districts of Bakali and Tyumazi near the border with Tatarstan. To a large extent they have tended to intermarry among themselves, and not with other Tatars or even with Russians. They also speak a form of the Tatar language which is often incomprehensible to ordinary (nominally Muslim) Tatars because of the absence of Arabic loan expressions and of Arabic-based words for biblical names.[16]

A Bashkort man playing a traditional Bashkort musical instrument

For the Muslim Tatars, religion had played a dominant role in the formation of a distinctive ethnic consciousness. During the 19th and early 20th centuries, Kazan was one of the four or five major centres of Islamic learning, attracting prominent Muslim scholars from all over the world and becoming one of the leading centres of Islamic revival in the world. Its Islamic university had about 12,000 students in 1913, including hundreds from many other Muslim countries. Every year more than 400 books were published in the city, their total circulation exceeding four million.[17] In 1906 a 'Muslim Union', also known as the 'Union of Russian Muslims', was founded in St Petersburg and eleven of its fifteen Central Committee members were Volga Tatars. However, the Communists also suppressed virtually all public expressions of religiosity, whether Islamic, Christian or other; in Kazan the Marjani mosque was the only one left functioning during the Soviet period.

Fear is often used to control an oppressed or repressed ethnic group: it is

The Marjani mosque in Kazan

not necessarily the use of violence as the threat or fear of it. This psychology has been maintained and perhaps even intensified by the Soviet police state and a fear of informers. Even in post-Soviet times, a Tatar friend of mine was afraid of having her name or information quoted by me when she confided in me some of the ways in which Islamic mullahs had been trying to persuade Tatars to attend mosques or to fast during Ramadan.

Influences on Christianity

In Siberia it is likely that widespread concerns with safety and security date back as far as human settlement in the region, giving rise to religious expressions of these fears. What is not so clear is whether or not in Central Asia fears about the evil eye and other such concerns predated the advent of Islam. Certainly fears of the evil eye are widespread in Muslim areas, but it is an aspect of folk Islam based on the *hadith* ('sayings' of Muhammed) with no direct mention of the evil eye in the Qur'an itself.[18]

A question then arises whether or not forms of Christianity in Russia have also been influenced by a culture in which fear has been a powerful influence on many people. I would suggest that such a process has been at work, perhaps even to some extent unconsciously, through the ways in which certain elements of Christian teaching have been emphasised to a greater extent than others.[19]

The ancestors of many Udmurts, 'Christianised' Tatars and other peoples were forced to convert to Russian Orthodoxy through coercion and even sometimes threats to their lives: therefore fear lay at the very root of their 'Christianity'. What impression might this have given them about the nature of the Christian God? Would he appear to them as a loving heavenly father, or — more likely — as a tyrant who forces his will on them, like the Russian Tsars? Even some Russian Orthodox Christians suffered persecution and harassment in the 17th century when they disagreed with the reforms initiated by Patriarch Nikon, who has been described as having had "an extraordinarily cruel nature" so that he "began to introduce uniformity with extremely cruel measures."[20] If those who suffered repression by the clerical authorities began to form a basic attitude towards God as similar to a despot, their descendants could also become socialised by example into holding similar kinds of attitudes. Why, for example, should an Udmurt pensioner, who, as far as I am aware, is not a practising Christian, write to me that she thought her difficulties in life (economic hardships and experience of widowhood) were punishments from God? Could it be anything to do with a sense of oppression within the history and psychology of the Udmurt people? Nowadays a common stereotype of the Udmurts is that they are peaceable and passive people, but this is largely because they have become accustomed to turning their anger inwards instead of expressing it on the surface towards others.

Several Udmurts have commented that fear seems to have a powerful influence on their own people. For example, a different Udmurt pensioner noticed how an Udmurt man was apparently uncomfortable and even (in her eyes) "afraid" of speaking with me when I addressed him in Udmurt after a church service. The pensioner thought it was not so much because I was a foreigner or had addressed him in Udmurt but because "he is a typical Udmurt and a lot of Udmurts are like that."

A perception of God as someone to be feared has apparently established itself also among a substantial number of Russians. For instance, in conversation with an Orthodox woman named Tanya in Yekaterinburg I was surprised at how often she referred to scriptures such as "the fear of the Lord is the beginning of wisdom."[21] Although she readily acknowledged the love of God too, it appeared to me as if the concept of the fear of God was somehow more deeply rooted in her consciousness.

On another occasion, when I had the opportunity to travel to the area around Nizhny Tagil, Tanya suggested it would be interesting for me to meet with an Orthodox priest named Father Michael whom she had telephoned so that he would be expecting me. However, when I arrived at the church I was met by a woman who said that Father Michael had gone a long distance away, to Verkhatoury. Whether or not she had been instructed to deceive me I cannot tell, but shortly afterwards I returned to the church and saw her weeping before an icon and begging God to be merciful and forgive her. We can only speculate what feelings and motivations were inside her, but my impression was that there was a sense of fear lest God should punish her.

After Atheism

Much more explicit was the attitude of some Russian Pentecostal Christians at the end of a meeting at which they had expected miraculous healings but apparently none had occurred. Then the leader of the meeting began to tell those present that God was punishing them. A former member of that church explained to me that this was a very common theme among Russian Pentecostals. Those who have suffered persecution for their religious faith could easily develop also a 'persecution complex' with regard to God, by regarding their earthly sufferings as being punishments from God.

Perhaps Russians too have a legacy from the very beginning of Russian Orthodoxy, when the ancestors of the Ukrainians and Russians were converted to Christianity largely by the dictates of their rulers, notably the decree of Prince Vladimir of Kiev in AD 988.[22] Since that time the Russian Orthodox faith has been a tool of politics to a greater or lesser extent, so it is not surprising if some Russian Christians should have developed an image of God to some extent coloured by perceptions of the Tsars.

One might expect such a perception to have been removed by seventy years of atheistic Communism, without the 'role model' provided by the Tsars or the former aristocracy. In spite of this, it seems as if both Orthodox and Protestant Christians are influenced by the wider culture in which there has been a long history of rule by force and coercion. The Communist party inflicted its own version of totalitarianism upon the people, and replaced one form of authoritarianism and despotism by another. Over the centuries, whether ruled by Tsars or the Communist party (or having to pay tribute to the Mongol conquerors or to the Tatar Khanate) political and social life has been dominated by a climate of repression and fear. Under such circumstances it is not surprising if a sense of fear should influence the religious conduct of ordinary people too.

The erosion of trust and of love between people cannot be blamed solely on the effects of an atheistic ideology. Many other factors have been at work, such as the expansion of big cities in which people become more anonymous and often feel lonely, unable to trust their neighbour in the next apartment. This was one legacy of 20th century life, but was partly reinforced by the erosion of religious values during the era of Socialism.

Fear breeds suspicion and distrust of others. Often in Russia people distrust those whom they suspect of lying to them or deceiving them. Nowadays there is often distrust in economic dealings, not only on account of the powerful influence of criminal elements — the 'Mafia' — but also because many people in their everyday experiences of life have become accustomed to being suspicious and distrustful. The only remedy to distrust is to build bridges of trust. Trust is usually built up over time by experience of a person, but it is also helped by an understanding of the other person's values and ethics. Atheism did little to provide a moral framework for life, and now people are looking for stable foundations for morality on which trust can begin to be built. Historically this has been a role played by religions. Even though

Russian Orthodoxy was exploited by the State for its own political purposes, its holy book nevertheless contains the statement: "There is no fear in love. But perfect love drives out fear, because fear has to do with punishment. The man who fears is not made perfect in love."[23]

One common fear is the fear of death — the 'great unknown' which all of us must face at some time. It raises religious questions about whether or not there is an afterlife. As such, it is intermediate between two of the levels of religiosity mentioned in my introduction. Whereas the dreams and visions discussed in chapter two clearly belong to the level of 'spiritual experiences', death belongs to the level which I described as "issues of life which might evoke religious questions." On the other hand, some spiritual experiences — such as 'near death experiences' or reports of apparitions identified as those of deceased people — are closely related to the events surrounding death. Cultural responses to death in the form of ways in which people remember those who have died also take religious forms in many societies. These are issues which will be discussed further in chapter four.

4
Death

"*...I* am writing to you now in greatest need and desperation. It is difficult for me to write or think clearly, but I hope you'll understand and excuse me. My son . . . died a month ago today. It was no accident, he did it gladly and willingly himself. He went up to the top of a TV tower . . . and stepped off into the sky."

In such terms a Russian mother described to me the suicide of her teenage son, her only child. She went on to write: "He was very much interested in religion, and it turns out that for him it was very much more than religion, or interest, or anything there's a word for. He did not believe, he knew there was life after death, he was sure this life we live here is something of a practical joke; he felt he didn't belong here, he sought desperately for the meaning of life, eternity, "home", where his true self, his spirit, came from and where it must eventually return. And he wrote in his last note he couldn't wait any longer.

"You know how rational and materialistic I used to be. I'm not any more. But I haven't become yet anything else. There's just no more me. You know, I'd told you more than once that I'm a mother first and only after that am I anything else. I can't stop being a mother overnight. So I live now in a state of an earthquake that never ends, without any ground under my feet. But I have one last duty for my son. I have to understand him, I have to find him wherever he is. I do not need any consolation, I can't be consoled. I'm very grave and serious in trying to find out whether there's God, or soul, or spirit, whether there's life after death. I'm very, very desperate to find out. Otherwise, if I will have to admit he made a mistake, if there's nothing beyond, if it is the end, then I myself . . . have no right whatever to go on living . . ."

This mother's distress had been compounded by the attitude of the Orthodox church in her city, which, regarding suicide as a serious sin, had refused to perform a funeral for her son. When I met the mother again some months later, in September 1994, she had become disillusioned with institutionalised religion, but she was still seeking answers to her fundamental questions about the meaning of life and whether or not there is a life after death. Probably her case is representative of many who, following the death of a close relative or friend — especially in circumstances such as this — have begun to ask questions of a spiritual nature. Is there really a life after death? If so, is there also a judgement by God? Even those who appear to be agnostics

Death

or atheists often begin to consider such questions when confronted by the realities of death.

The experience of having come very close to dying often prompts people to re-examine their values and priorities in life. Some experiences reported by those who had clinically 'died' but who later recovered will be related in chapter 12. What might be interpreted as a possible 'near-death experience' was described to me as a "dream" by a Russian pensioner whom I met on a train journey in October 1998. She said: "About ten years ago I was very seriously ill and had to have an operation. However, the doctors and medical staff thought I had died. They said that they couldn't save me. At that time I had a dream in which I saw my dead mother come and sit on the end of the bed and invite me to go with her, but I didn't go."

This woman was about 55 years old at the time of her experience, having retired from her teaching job shortly beforehand. In further discussion with her, she mentioned the changes which had taken place in her life since that time and which appear to have been at least partly a result of her close encounter with death:

"Since that time I have become a completely different person from the one I used to be. Other people, including my husband and daughter, have noticed it too. I'm now kinder, more gentle, and I don't argue or get angry any more. I've also started to write poems — about love, nature, and also some sad ones about that serious illness. Shortly after my time in hospital a friend of mine arranged for me to meet an Orthodox priest. She knew I was shy about going to church, as I'd been brought up as an atheist, so she invited me to her home when the priest was there too and left us to talk. The priest has since given me various books. I also bought a headscarf and have started to go along to the church."

Her close encounter with death has clearly influenced some of this woman's attitudes to life. Her values and character have altered to some extent, and she has begun to express herself in a creative way, by starting to write poetry — two examples of which she showed to me. These are expressive of a kind of 'spirituality' but not of a formal religious type: in fact she remains critical of those who go to church and confess their sins but still live their lives in the same way as before. She also has a strong belief in astrology which has apparently not been affected by the priest's negative attitude to astrology as incompatible with a Christian faith.

However, the economic turmoil in Russia since the collapse of Communism appears to have precipitated a rise in suicides among those who felt they had no hope in this life. Similar rises occurred in Kazakhstan, particularly among men, and probably elsewhere in the CIS too.[1] However, the highest incidence of suicide within the Russian Federation occurs not among the Russians but among some of the indigenous peoples of the Urals region and Siberia. It is particularly high among the Udmurts and other peoples speaking Finno-Ugric languages.[2]

In Udmurtia I was told suicides had been not uncommon among young men in some of the Udmurt villages. The reason given was the fact that many young women had migrated to the cities while the young men had been left with responsibility for the farms. Having little hope of finding a partner in marriage, they had decided that life was not worth living.

Several Udmurts commented to me that the relatively high suicide rate is probably connected with their national character. They say: "We tend to keep our feelings inside ourselves and to bottle up negative emotions like frustration or anger. On the outside we seem calm but there is no outlet for our negative

Udmurt musicians

feelings. Then, when things seem too difficult to cope with, suicide might seem to be a way out." A reticence about expressing inner feelings can affect family relationships in other ways. For instance, the wife of an Udmurt man told me that she often senses when things are not right with her husband but often he is reluctant to express even to his wife about his frustrations at work.

An Udmurt woman who accompanied me to a service at a Pentecostal church in Izhevsk was disturbed by the public weeping which took place when the Holy Spirit was invited to minister to people. She later explained that she felt uneasy about it because Udmurts do not express such deep emotions in public. However, she also recognised the importance of allowing such emotions to be released in an appropriate way. In that context, it was believed that God wanted to release the pain of the deep hurts which had been bottled up inside.

The 747,000 Udmurts live principally between the Vyatka and Kama rivers, a little to the west of the Urals. To their west are the Mari and Mordvin, other Finno-Ugric peoples with populations of 671,000 and 1,154,000 respectively. The ancestors of other Finno-Ugric peoples appear to have spread out from an original homeland in what are now the Mordvin,

Mari and Udmurt regions, from which the Khanty and Mansi went north and east across the Ural mountains, the Komi to the north, Finns and Estonians to the north-west and Hungarians to the west. However, even within Europe, suicide rates are relatively high among the Finns and Hungarians. Therefore it has been suggested that economic problems might precipitate suicides, but deeper ethnic, cultural or spiritual (or, in religious terms, 'demonic') factors might have given these peoples an underlying disposition towards choosing suicide as a means of escape.

In Siberia, there are 'unexplained' cases of suicide not only among Finno-Ugric peoples such as the Khanty and Mansi but also among other indigenous peoples. Recently an Evenki man, out fishing with his son on an island in a lake, told his son to wait where he was while the father disappeared among some trees. When the father failed to return after a while, the son went to look for him and found his father had hung himself by a rope from one of the trees. The distraught family were unable to explain the suicide.

Suicides in contemporary Russia could be symptomatic of many different kinds of social and economic malaise. For some of the indigenous peoples, they might be results of the collapse of a traditional lifestyle on account of Soviet Russification policies. More recently the collapse of Communism has led to economic chaos and uncertainties about the future. Other factors, such as a shortage of marriageable women in some villages, or personal failures of one kind or another, are all potential contributory factors. However, the major influence appears to be a lack of hope. It is the same influence which often leads people into alcoholism or other forms of 'escape'. However, those who have finally concluded that life is either meaningless or not worth living might give up hope and choose to end their own lives. As such, suicide is a response also to questions of meaning in life which are religious issues.

Certain spiritual experiences occurring at the same time as the death of someone else are likely to be interpreted as being more than coincidence. The following account from an elderly Russian woman could be interpreted as being an attack either from the spirit of the dead person or by an evil spirit associated with that person seeking a new abode: "One night I had a vivid dream in which I was being attacked by a kind of vampire or evil spirit. Even though I was an atheist at the time, in my dream I began to recite the Lord's prayer and the 'vampire' departed. I later discovered that on that very same night the daughter of the woman in the apartment below mine had died. My instincts told me that she had died from an attempted abortion, though I don't know why she didn't go to a hospital for it. Perhaps it was because she had married a Central Asian man who had probably been released from one of the penal colonies in our region. He wanted lots of children but his wife didn't want an additional child so I suspect that's why she chose to have a secret abortion. At the time of this dream I was an atheist so I don't know why a friend gave me an icon when I went in to work the following day after that horrible dream. Now I am more inclined to think that there is a God."

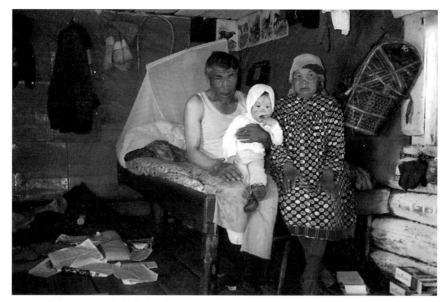

A Mansi couple with their grandson in their home downriver from Lombovozh. The spirit of a dead relative was believed to have come into this child.

Generational spirits

Several other questions of a religious nature are also raised by experiences of death. In some parts of Siberia they are fuelled also by accounts of spiritual experiences involving what are believed to be spirits of departed ancestors. For instance, in north-west Siberia an older Mansi couple living by themselves in the forest had their grandson staying with them when I visited their hut. Referring to the child, the old man said: "That's my grandfather." When the child was born, it was discerned through divination that the spirit of this relative had entered again into the baby, so they refer to the child by a kinship term referring to someone of that previous generation.[3]

Informants were not always very clear about the nature of this divination. Partly this was because it was performed by older women who are no longer alive. The clearest description I obtained was as follows: "My clan used a kitten to find out which soul, of which relative, had entered a newborn baby. The kittens were never to be hurt or killed and were sacred animals in their clan. Ours was the only family which used kittens for this process: other families might do it in some other way.

"Special women wrapped the kitten in a piece of cloth that had been used in childbirth to wrap the newborn baby, or one which had been in contact with the infant. It was believed that a tie or line of communication was established between that kitten and the things stored in the attic — as the belongings of people who have died are usually stored in every home. Then

Death

they lifted up the kitten, so wrapped, several times. In the process they were thinking of one or another of the dead relatives. If the kitten was light and they lifted it easily, it meant that there was no tie between a particular dead relative and the newborn baby, but when they felt that the weight was heavy it meant that it was the soul of that person who had been born again in the baby."

Belief in the transmission of generational spirits is also quite strong among the Yukagirs who live on the tundra near the Kolyma river in northeastern Siberia. Several Yukagirs related to me experiences in which a young child might refer to an adult in a manner similar to that of a deceased relative. This was taken as a sign that the spirit of a dead relative had passed into that child.[4] For instance, in his childhood a young man named Igor saw a girl whom he had not met before but he referred to her by a term of endearment which had only been used by her dead father, who was Igor's grandfather's brother. The family took this as a sign that the spirit of that relative had come into Igor.

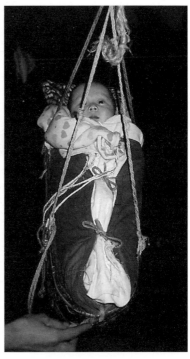

A Mansi baby

When Igor's brother, Nikolai, was a child he came out with an account of how he had come out of a bar in Cherskiy while drunk and had fallen over in the street. He had felt very cold and had died of the cold. Relatives in Cherskiy knew of a man who had died in this way but the family claimed that Nikolai knew nothing about this.[5]

Igor's sister, Nadia, related how, when she was about three to five years old, she had looked out of a window and said "Those children are all my children"; in fact, they were adults in their

A Yukagir reindeer herder

thirties but were all children of Nadia's grandmother's brother. Nadia went on to speak of the occasion when a girl came to visit them while they were living in Cherskiy. At that time Nadia's son was aged about three or four and did not know the girl but he said "my daughter is coming" in a manner reminiscent of the girl's father.

The interpretation of such accounts is influenced to some extent by one's cultural background and by one's personal choice of what one wants to believe. Some might see them as merely constructions of the imagination, reading more than is warranted into childish comments. Another interpretation is to see them as reconstructions based on details learned at different times and buried in one's subconscious memory. The Yukagirs choose to interpret these experiences as evidence of the transmission of generational spirits: this is not dissimilar to the process among many Siberian peoples whereby the person whom the spirits call to take up the role of shaman is often (but not invariably) a member of a family in which there had been shamans in previous generations.[6] All these explanations are to some extent influenced by one's cultural background and by personal choice regarding viewpoints which are felt to be acceptable. Therefore a Yukagir woman remarked that conversations about the transmission of generational spirits are restricted to the Yukagir and Even people because "Russians do not speak about this."[7]

A Yukagir man named Nikolai (the brother of Igor and Nadia) with the tooth of a mammoth

What might initially seem surprising is that a belief in the transmission of spirits from one generation to another is found not only in Siberia but even among some nominally Muslim Turkmen families. They believe that some ancestral spirits can come back again, but always in the male line. For instance, when a baby boy was born on the fortieth day after the death of his grandfather it was believed that the spirit of the dead ancestor had passed into the child. It was felt to be more than 'coincidence' that the birth took place on a day for remembering the deceased ancestor. Therefore those in the family often referred to the child using the kinship terminology which they would have used towards the dead man. If necessary an older person could rebuke the child for misbehaviour but in general family members would behave with respect and deference towards the child. In other instances a child might be born several years after the death of a relative but be named after the dead person. If the child then shows behaviour patterns similar to those of the dead person, the relatives would take this as a confirmation that the spirit of the dead person had come into the child and would refer to the child in terms

Death

Mansi people mourning besides Domna's body.
(The photo was taken at their request as a memorial to her.)

appropriate to the deceased person. A Turkmen man whose younger sister had died subsequently named one of his daughters after the deceased sister and his behaviour towards that daughter was reported by others in the family to be more deferential and respectful than towards his other children. Although my informants believed such beliefs to be quite common among the Turkmen, others had not heard of such ideas. In a supposedly 'Muslim' cultural context such beliefs are certainly not 'orthodox'.

Remembering the dead

What is more common to many different peoples, however, is a felt need to remember the dead in some way, as illustrated by the following case studies.

Mansi and Khanty

In her one-room shack, Domna's body lay on the bed, covered by a blanket. Her body had been found that morning at the foot of a steep river bank behind her home. A few local Mansi women arrived to pay their last respects and one of them began to cover over the mirror on the wall. Once a pathologist had flown in by helicopter from the district administrative centre and had examined the body, the local people were allowed to prepare for the funeral. Domna's body was dressed and laid out in an open bier in the centre of her living room. Her face was left visible but her body was covered over with blankets. One of the women

After Atheism

Khanty graves, with cups left in front after relatives had eaten and drunk in memory of the deceased

told me that certain cloths put in with the blankets were supposed to help her spirit in its transition to the other world. For the next three days until her funeral the body was always constantly attended by Domna's relatives and friends in the village. Visitors would come and sit there with the relatives as a sign of respect to the dead person. Her two older children were brought up by boat from the village where they had to attend a boarding school, as there was no school in this remote Mansi village. They too sat with their mother, sobbing their eyes out.

Under Russian influence, corpses are now buried underground, but traditionally were left in wooden 'coffins' above ground. In Tsarist times some of the Khanty and Mansi had become superficially Christianised, so nowadays their graves have crosses on them, but in practice many pre-Christian traditions have survived. Nenet graves generally lack crosses because they were not Christianised like the Khanty: modern Nenet graves have a fence around them, within which may be placed the dead person's sledge or other personal possessions. These are broken, so that they too are 'dead' and can pass into the next life with the owner. Similar customs of damaging objects left for the dead spirit are reported also for the Evenki and Even people of Eastern Siberia.[8]

Among the Khanty and Mansi, valuable items such as guns, fishing rods or clothes which might be stolen from the graves are nowadays wrapped in cloth and stored at home in a special trunk. This is kept in the attic or in a grain store where household gods are kept and to which women are not allowed access. Since personal items such as clothes can serve as means of

Death

A Mansi grave, showing cups and the 'stopper' by means of which offerings are placed inside the grave cover

communication with the dead, they are taken from the trunk into the forest when the living try to make contact with spirits of the dead.

One young Khant man said to me: "I remember my dead relatives, either at the graveyard or else at home, when I dream about them more often than I feel is usual. Those dreams mean they have lost communication with me and they are in need of such communication. To restore that bond, I go to one of the sacred places and bring there something which belonged to that relative which he had used and liked very much; it might only be something small. I'd also bring some food and drink with me, pour out a little for the spirit, eat and drink there and then tie a piece of white cloth on the branch of a tree. While in that place I would communicate with the dead. After I leave, that piece of white cloth would allow the dead to restore the connection: they would know about my life and be able to communicate."

Mansi people visit the graves once or twice a year to eat and drink in memory of those who had died. They might also place food or drink on the graves as offerings for the spirits of the dead. Among the Khanty, kettles and saucepans for this purpose can often be seen in front of the wooden structures above ground which mark the sites of the burials. Mansi graves are a little different, with a wooden 'stopper' at the front of the structure. The body is buried under the ground but the structure on the surface has a stopper which is removed so that food and drink offerings can be placed inside.

A Mansi woman named Anna says that she goes to the graves of her dead relatives twice a year, in the spring and autumn, but if a person has died recently she goes more often. At the grave she makes tea and prepares food in memory of the dead. Iliya, a 40-year-old Mansi man, says that he visits his parents' graves on the 9th of May — Victory Day — because his father took part in the war. He brings with him some food and alcohol (if any is available). When Iliya comes to 'greet' the dead, he takes away the stopper of the wooden grave cover and he knocks three times on the corner "to greet my parents and tell them that we have come." Then the living relatives eat and drink, sitting around in a circle and talking for an hour or two. They remember the dead, especially talking about the dead person's good qualities. When they leave, they put the stopper back and leave some water in a cup inside the cover.

"Fifteen or twenty years ago," says Iliya, "when it was easier to obtain alcohol, they would leave a fresh bottle of vodka and take the old one out to drink it. Nobody would steal the vodka in between the family's visits to the graves. Nowadays, however, alcohol has become a luxury and cannot be thrown away." Iliya goes to remember the dead but he also has some belief in a life after death. From his mother he had heard that the dead person's soul goes to the north, to the island of Novaya Zemlya. According to Mansi tradition, as remembered by Iliya, a man has "four or more" souls; other people say that a man has five and a woman four souls. Therefore one of the souls is permanently left at the grave site.

Others made the same kinds of comments, sometimes with small variations. The family of a 67-year-old Mansi man named Igor used to put wine at the graves but did not drink the old wine on their next visit. Instead they would leave the old bottles where they were and add new ones. However, they do not leave food because the dogs would eat it.

Udmurts

An Udmurt woman in traditional dress

Among the Udmurts, memorial customs for the ancestors are deemed by many families to be very important. Various local traditions say that the souls of the dead have gone under the river or else to the lower reaches of the river. In some villages a memorial ritual is performed in the winter but prior to the winter solstice, before the position of the sun begins to come northwards once again. They believe that they should always feed the spirits of the dead by the men sacrificing a pony and the women a cow. If the ritual is not performed, it is thought that the ancestors would prevent childbirths from taking place in the village, because the spirits of the dead are said to have power over the fortunes of childbirth.

Death

Many Udmurts believe that the ancestors come to them in dreams. Some say that if they see a dead person in a dream it is an omen that misfortune will occur. Sometimes if a person becomes ill, others might say the illness is a punishment on account of that person's neglecting to remember his or her ancestors. Other memorial rites also take place in the spring and autumn, when a married woman whose parents have died might sacrifice a pony in memory of her own parents. She has to do this at least once in her lifetime, at least one year after the death of her parents. Traditionally it had to be a pony she had reared herself, not one she had bought, but nowadays this rule is relaxed to allow not only purchased ponies but even cows to be sacrificed instead. The significance of the rite is related to the fact that a married woman has left her natal home and married into another family, but still retains a responsibility towards her own parents.[9]

The memorial to a deceased relative in an urban Udmurt home

Both in towns and the countryside some Udmurt families put aside food for the dead on the occasions of the spring and autumn memorials. The food put at the graveside or sometimes in the family's garden is eaten by the birds and wild animals but was traditionally regarded as food for the dead: there is at least a 'feeling', if not a 'belief', that the soul of the dead person had not completely gone away and could receive some nourishment from the memorial food left out both in the home and at the graveside. After a person dies there are special memorials on the first, ninth and fortieth nights after the death and on the first and third anniversaries of the death, in addition to the general memorials in the spring and autumn. However, during the first forty days practically every time a family has a meal they may also be thinking about the dead person while they eat.

An urban Udmurt family I visited a year after the wife's father's death had put out on the table, as special memorial food, rice with raisins mixed into it and also a dish of honey. At the beginning of the meal the wife poured out some wine into a glass which was then passed around those present, each person in turn taking a sip and then passing on the glass. On a separate table, in front of a photograph of the deceased, were placed a vase of flowers, a candle and a small plate on which was some pie, a slice of bread and some confectionery (which are later eaten by family members). This urban family, which has also been influenced by Christianity, are less inclined to believe that the food nourishes the dead soul: instead, they regard this custom as "just a ceremony" which is a sign of respect for the dead and an expression of remembrance.

Left: Udmurts paying their respects to the soul of the dead man.
Right: Placing offerings in the barrel containing the skull of a horse.

Even though many Udmurts also attend Orthodox churches, they also perform these traditional rites for the ancestors. Families who have suffered a death within the previous year are regarded as in some sense polluted and so do not kneel to pray with others in an Orthodox church. The wife in another urban Udmurt family told me that every Saturday she recites the names of all the dead they can remember in their families on both the father's and mother's side: she claimed to know the names of seventy-seven generations. This family pass around a glass of brandy in an anti-clockwise direction and some cake or bread in a clockwise direction — explained as being "in the direction of the sun." All those present partake of the bread or cake, but only the adults drank the brandy.

The Udmurts say that a person can undergo three 'weddings': the first is at birth, and is a wedding to this world, the second takes place at marriage, and the third is after death — as a wedding to the other world. The spirits of ancestors are 'not revered or worshipped' but are 'communed with and cajoled' so that they will not drag the living into the next world. After death the spirits marry again, so they must be sent a dowry two or three years after the person's death.[10] Rites connected with this third wedding are condemned by the Russian Orthodox church (to which many Udmurts also belong) because it is viewed as a kind of occultism which can make a person susceptible to demonic influences. In 1994 I witnessed in the southern

Death

Udmurt village of Bolshiye Siby the re-enactment of such a pagan rite which was held on behalf of a man who had died, in order to help him to marry again in the next world.[11] For this ritual a group of local people — mainly older women, plus a few men and children — dressed up in traditional costume from their district of southern Udmurtia. The women wore red shawls, dresses painstakingly embroidered with traditional designs and on their chests wore rows of coins, many of them dating back fifty to one hundred years. On their feet they wore the plaited birch bark sandals which had been widely worn in country areas until the 19th century. Such sandals were also worn by some of the men, whose other clothing consisted of more modern style shirts or jerkins.

Above: Transporting the barrel on a sledge to the edge of the village.
Below: Burning the barrel, skull and offerings

After Atheism

Eating and drinking together after the ceremony

Prior to the commencement of the ritual, people gather at the home of the deceased man and begin to drink vodka together, offering glasses to one another to drink. Then the skull of a horse — representing the dead man — is brought out and placed in a kind of wooden barrel. On top of it and around it are then placed in turn various offerings of food, cloth or coins, expressing the idea that the person would need such things in the next world in order to marry again. The living wish the deceased to have contentment and a good marriage in the next world.

After this, the participants change their dress, the women putting on men's hats and coats, and the men putting on women's headscarves and skirts over their trousers. This ceremonial transvestitism is supposed to confuse evil spirits about the identity of the participants. The barrel containing the horse skull and offerings is then dragged on a sledge over the snow to a tall evergreen tree, symbolising the passage from this world to the next. There the offerings and skull are burned, the smoke rising up in the air next to the tree as a kind of vehicle for conveying the offerings into the other world.

As the fire burns, people share together in eating a kind of pastry and drinking vodka in memory of the deceased. Once the fire dies down and is doused, the children are encouraged to look among the ashes for the coins, which they retrieve. Finally, prayers are said — first to the pagan gods and then in the Russian Orthodox manner to the Christian God. The participants then return to the house for a meal together and further mutual sharing of drinks.

Death

Mongolians

The death of a relative can often act as a prompt towards religious behaviour among those who would not normally consider themselves to be religious. For example, a 45-year-old Mongolian woman said: "I was one of five children but my younger sister died in 1985 at the age of 25. My mother sent someone to Gandan monastery after the accident in which my sister died: before that, my family never went to Gandan. It was only after my sister died, not before, that my mother and I began to go to Gandan. I wanted to ask God to help me but I'm not a believer.

We put a *khadag* (scarf used for offering a gift), food and a candle in front of a picture of my dead sister. There is a ritual to do this for 45 days after the death: my mother did this and the lamas read sutras. We paid for them to read the book necessary for my sister. Every day we put food and drink (milk or tea) in front of the photo: some people put food for the dead before the photo every year too, for a few days after New Year. Sometimes (such as at New Year) my mother lights a candle in front of my sister's photograph: I don't know the meaning of the candle but I think it is because the way my sister has gone is very dark and she needs a candle to light the way. I don't know where my sister has gone. It's a great loss for my family, but I don't feel anything to do with an afterlife. I don't think my mother believes in an afterlife either: my parents don't believe in God and never speak about God or do religious rituals."

Gandan monastery, Ulaan Baator, Mongolia

After Atheism

Other Mongolians also mentioned visits to Gandan monastery after the death of a parent but saw it as a 'tradition' rather than as motivated by any belief. A 40-year-old Mongolian man, for instance, said that they keep a photograph of his deceased father but do not put any candles or food offerings in front of it. It seems as if more 'religious' behaviour is sometimes prompted by religious experiences such as that of the 41-year-old Mongolian woman who visited a lama after she began frequently to dream about her dead mother (see p57). However, she had also had contact with the monastery to arrange the rites performed in her mother's memory: "After my mother's death we went to a Buddhist temple and felt comforted. My mother's eldest brother helped me as it was my first experience of this; we gave some money. Then we went to the lama and asked for the traditional burial rites and arranged the date and time. We asked what we should put in the grave and the lama told us to put in what the dead person wanted us to do. I'm not a believer but what was most interesting to me was that he asked about my mother's death and her date of birth which he looked up in books. He said my mother's spirit was asking me to devote to her a 'valuable thing'. I knew she wasn't interested in gold or money but then I remembered that at the time of my mother's death she had wanted us to buy a pain-killing medicine which contains many types of precious minerals including gold. She had been waiting for this and we had found it from the Indian embassy. Then we read that it can't be taken with alcohol and we were afraid of the reaction but the only possible way to swallow it was with vodka. My mother had said she'd not take it because people would say we'd killed her. She died a day later. So I was surprised when the lama spoke about this, so we put the medicine in the soil under her grave.

"After my mother's death we were told to make a portrait of the 'White Goddess' — the deity who gives happiness to women especially. Her portrait was done for us by a painter and we now keep it in our apartment in Ulaan Baator. For forty-nine days after her death we also had a time of mourning but on the forty-ninth day we invited in the children with the relatives and we gave something — food or sweets — to the children and also to the dogs, but especially we wanted to make the children happy on this day. We also lit a candle

Kalmyks, who, like other Mongolians, perform rites in memory of the dead

or an oil lamp in front of her portrait, in memory of the dead person, at the funeral, and then after three days, forty-nine days and annually on the anniversary of the death. In front of her portrait we also placed some sweets and small dishes of food from our dinner, which we gave to our mother.

"Once or twice a year I go to the place where she is buried. There I can't say anything aloud because others are with me but inside I say 'help me!' or 'help my father, or my children, and so on!' I think that others do the same in such places. Sometimes in my daily life I say 'help me, mother!' just like other people say 'oh God, help me!' "

Another instance which might be regarded as communication through dreams with the spirit of a dead person was recounted by a 30-year-old man, who said: "I've often dreamed about my father. In my dream I talk with him or I'm sitting and doing something with him, like we used to do when he was alive. I don't tell anyone about it, but I often dream about it. When I was studying in Moscow, he died. After that I was very unhappy because I couldn't see him again. I thought about him very much and about the ways I'd do various things with him if he were alive. This feeling was fixed in my soul. Sometimes I don't believe he died without seeing me. It's difficult for children when a parent dies. I dream especially about the times before he died: I never dream of present circumstances. However, I don't feel his spirit comes back when I dream of him: there is a difference between the spirit of a dead person and a dream about a dead person."

A dream prompted a 34-year-old Kalmyk man to visit a cemetery and perform rites. He said: "Kalmyks say that if you see yourself dead in a dream, or dream of dead relatives calling you, it means that you have to go to a cemetery to remember them and drink some vodka — the dream means their soul is missing you. I did that when I dreamed of my dead grandfather calling me. I also go there every year in August on the anniversary of the death of my friend who drowned three years ago. We remember the dead person, drink a little and leave a glass of vodka at the grave, along with some sweets and biscuits. It's difficult to say what my feelings are at the time; each feels his own feelings. For me, I feel that I'm not alone, that I have friends and I think that I'll be remembered after death too."

Russians and Yukagirs

Practices such as this are reminiscent of East Asian ancestral rites but also within Russian Orthodoxy there are rituals for remembering the dead. (The Kalmyk man just quoted speculated whether the Kalmyk practice was possibly borrowed from the Russians.) It is not uncommon in some areas of Russia for people to visit the graves of dead relatives and there put offerings of food or drink as well as flowers. While passing a graveyard in northern Siberia, I heard a woman weeping and crying out: "Why did you leave me?" At first I wondered if she were from one of the indigenous peoples but Natasha turned out to be Russian.

Her mother had drowned in a lake. Natasha often visits the grave, where she and some other relatives were drinking vodka not only in memory of the dead but also to drown their own sorrows.

Sometimes the non-Russian peoples point to minor differences in the timing of their memorial rites as indicative of the distinctions between their cultures and that of Russians or other neighbouring peoples.[12] For instance, among the Yukagirs a dead body is laid out in the home for three days after the death and after that for a further three days in a kind of tent on the street, where at mealtimes people come to offer food for the dead person's spirit by burning food on a fire. Then on the seventh day the corpse is buried and at the same time a reindeer is sacrificed "to help the dead person to enter the next world."[13] This custom of putting the corpse on public display from the fourth to sixth days after the death is regarded as a distinguishing feature of Yukagir culture in contrast to Russian customs.

Another distinguishing feature was mentioned by a Yukagir man who commented that they go to visit the graves for three years after a person's death and then stop doing so, "whereas the Russians keep visiting for longer." A Sakha (Yakut) man said the same — that they also stop visiting after three years — except in the case of a powerful shaman. For a shaman the visits should be maintained for 300 years, in which time the tall structure on which the coffin stands has to be rebuilt three times. This is because a shaman has "a very strong and potentially dangerous spirit." However, he said, if a shaman does not want his spirit (or shamanic powers) to be passed on to his descendants then he asks to be cremated instead, so that "his spirit passes through the fire and becomes a dead spirit."

It appears that the practice of making offerings to the spirits of the dead has come into some branches of the Russian Orthodox church through the assimilation of pre-Christian practices such as those which have survived today among some of the indigenous peoples of Siberia and to some extent among European peoples such as the Udmurts and Mari. However, even within the Orthodox and Roman Catholic churches there had already been a tradition of prayers on behalf of the dead; according to Rogozin, such prayers developed in the sixth-seventh centuries AD.[14] In Russia, the conversion to Christianity of peoples who had formerly practised memorial rituals for the dead easily led to the assimilation of such practices into the Orthodox church, although some Orthodox priests remain opposed to grave offerings because they are vestiges of paganism.[15]

Central Asian peoples

A similar process appears to have taken place among the Kazaks and a number of other nominally 'Muslim' peoples of Central Asia. For instance, in June 1992 I came across a Kazak memorial rite held in a suburb of Almaty. Some Kazak acquaintances pointed out to me three yurts (traditional felt

Death

tents) which had been erected in a nearby courtyard in preparation for the memorial rites on the fortieth day after the death of a well-known local man. Some of his relatives who lived nearby had put up the yurts because they would not have enough room in their apartment

Cutting up the horse meat in preparation for the memorial rite

for all the visitors. When we visited the yurts we found that the relatives were in the process of cutting up a horse in preparation for the memorial rites to be held on the Saturday afternoon. The sister of the deceased man said I would be welcome to attend the events two days later.

On our arrival that Saturday we were invited into one of the yurts to eat with the family. After we had finished, subsequent visitors were invited to eat in the same yurt. An adjacent yurt was used for Kazaks to assemble while one of them recited Islamic prayers, while the third yurt (where they had been slaughtering the horse) was used as a kitchen. It appeared as if the whole day was occupied in entertaining guests who on arrival were entertained with food in the one tent and before departing would go to pray for the deceased in the other tent.

Among the Kazaks, usually a funeral is conducted no later than three days after a person's death: this is related to a belief that the soul will not be at peace until the body is finally laid to rest. However, there is also a belief that the soul remains in the proximity of the body for forty days before going either to paradise or hell. This is one reason for the memorial rites on the fortieth day after death, but other

Kazak prayers in a yurt on the fortieth day after a death

After Atheism

family gatherings to remember the dead also occur seven days after the death and on each anniversary of the death.

Funerals among Uzbeks are held as soon as possible after the death, often on the same day or the following day. A belief that the dead

A meal in a Turkmen home

person's soul remains for two days in the vicinity of the body appears to be related to a practice whereby for two days food is prepared not by the immediate relatives of the deceased person but by friends of the family. On the third day guests are invited to share pilau (a rice dish) served with the meat of a specially slaughtered sheep. Each Thursday for the first forty days after the death there is a gathering of the mourners to eat together and offer prayers. Similar memorials are held seven months and one year after the death.

Such customs relating to the dead are also found among Uighurs, Turkmen and other Central Asian peoples. For instance, Turkmen families often say that it is important to honour the dead in order to ensure one's own good fortune in the future. On the occasion of a wedding, Turkmen couples often place flowers on the graves of their deceased ancestors. To commemorate the anniversary of a person's death, Turkmen families often offer a small 'sacrifice' of food. In rural areas it might be a literal sacrifice of an animal but often nowadays the offering consists of baking bread which one distributes to one's neighbours or the purchase of sweets for children.

Turkmen women

A similar practice among Uzbeks is known as *khudai* and is seen as a ritual sacrifice of animals (in particular) designed to assist departed

Death

A Kazak 'folk' religious practice: handkerchiefs are attached to a tree as a form of 'prayer' and sign of respect near the memorial to Chokan Velikhanov, a 'father of the nation' figure. In many other locations in Central Asia similar pieces of cloth are tied to trees to express a wish or a prayer (originally to a local deity) for benefits such as safety in travel or the granting of a child to someone with no child, or perhaps with a daughter but no son.

relatives (or perhaps friends) to advance to a higher plane in the afterlife. Those who pray for (or to) the ancestors may be motivated not only by beliefs about the welfare of the dead but may also seek for themselves, in this life, the blessing and assistance of dead relatives.

The perception of many ordinary Kazaks and Uzbeks is that these memorial rituals are Islamic and therefore are important in themselves as expressions of their national identity in preserving their own culture and in distinguishing them from Russians. What is questionable, however, is whether or not these are truly 'Islamic' or are actually vestiges of shamanism and of pre-Islamic religious practices which have survived in many areas. There is little or no Qur'anic basis for memorial rites for the dead and in general among Muslim peoples "beliefs and activities involved in death rites lean heavily on the folk-Islamic view of the world."[16] Even though the Kazak participants themselves now believe that such practices are actually 'Islamic', it is more likely that they are relics of pre-Islamic folk religious practices.[17]

Memorial rites have persisted strongly throughout the Soviet period because of the strong cultural feelings attached to them. In Turkmenistan Muslim pilgrimages have even been made to the cemeteries of 'deserters' who, for religious reasons, had refused to serve in the Soviet army during the Second World War.[18] In such ways, apparently 'secular' sites became imbued with religious meanings.

Cultural values connected with memorialism

In all these cultures — most noticeably among Central Asian peoples such as Kazaks, Turkmen or Uzbeks, but also to a considerable extent among Siberian and European peoples like the Mansi and Udmurts — honour and respect is given to old people. I witnessed an example of this at a Mansi Christian meeting where those coming to the front for prayer deliberately lined up in age order, with the older people going first.[19]

This respect for older people also extends to those who have died, particularly those belonging to previous generations of one's family. One Mansi man expressed to me his view that the Mansi gods are actually ancestors of the Mansi people: a continuity between respect for old people and worship of the traditional gods is also shown by a traditional practice of offering sacrifices to 'effigies of the ancestors'.[20] Obviously similar motivations are at work in the lives of at least some participants in the memorial rites.[21]

Among the peoples of the north — as well as those of Central Asia — this respect for older people is one of the traditional cultural values which could

Top: An elderly Mansi woman. Bottom: A Mansi family

well be an important building block in the revival of these cultures within the contemporary post-Soviet context. Values of this kind are important in themselves and can also serve as a source of national self-esteem among peoples who for too long have felt themselves to be 'second-class citizens' in relation to the Russians. In giving respect to parents and older people, the values of the 'pagan' peoples of Siberia and of the Muslim peoples of Central Asia and the Caucasus are closer to the standards of the Bible than are the practices of many in the West or Russia who claim to be Christians.[22]

Nevertheless, the Khanty, Udmurts, Turkmen, Uzbeks and other such peoples, as well as the Russians, tend to place emphasis mainly on the formal rituals after a person is dead. It is far easier to perform socially expected rituals such as visiting a grave than to go to the trouble of visiting that person while he or she is alive, spending time with him or her, and caring for the person's human needs in this life.

Udmurts in traditional costume

Respect for parents and older people often develops not so much as an obligation but as a response among children who feel that they are valued and important. The Udmurts placed a high value on children, as shown by traditional proverbs such as: "A couple without children are orphans" (Нылпитэк кышнокартъёс — сиротаос / *Nylpitek kyshnokartyos — sirotaos*); and "A home with children is a cheerful one" (Корка нылпиен шулдыр / *Korka nylpien shuldyr*).[23] Corresponding to these are

proverbs about respecting parents, such as: "A person who esteems his father and mother will pass a century without seeing hardship" (Анай-атайзэ санлась мурт шугез адӟытэк даурзэ орчытоз / *Anay-atayze sanlas' murt shugez adjytek daurze orchytoz*); or "Even if you are living very well, don't forget your mother and father" (Туж умой ке но улӥськод, мумыдэ-айыдэ эн вунэты / *Tuzh utoy ke no ulis'kod, mumyde-aiyde en vunety*).²⁴

A high cultural value accorded to the rearing of children can be expressed in terms of quality rather than quantity, in so far as few Udmurt families have more than one or two children. Among peoples of Central Asia and the Caucasus, especially in rural areas, there is a greater emphasis placed on the quantity of children and on having at least one male offspring, but in recent years there has been a greater use of contraception to reduce family size. Among the Chukchi and Koryaks of the Russian Far East, a preference for no more than three children has been expressed by those aged less than thirty, whereas those older than thirty often preferred to have larger families.²⁵

A Kazak child

It is likely that the placing of a high value on having children also influences the prevalence of rites to give thanks for the birth of children or, among Muslim peoples, the persistence of circumcision as a ritual with strong cultural (rather than necessarily religious) symbolism.²⁶ In many cultures the timing of rituals after birth follow a similar and parallel pattern to those after a person's death: among the Kazaks, for instance, a newborn child is shown in public for the first time forty days after birth (having until then been kept from public view partly for fear of the evil eye). This rite (called *besik toi*) parallels the public memorial which occurs forty days after a person's death. It is noteworthy that no such forty-day celebration occurs among the Hui (Chinese Muslims) in China but those living in the former Soviet Union (where they are known as Dungans) now hold a celebration at that time too.²⁷ Within any particular culture, the significance of particular dates (in terms of the days after a birth or death when the event takes place) is dependent upon local

Chukchi children

concepts associated with those numbers. Among the Khanty of Siberia, for instance, the timing of memorial ceremonies varied but was generally "ordered by the numbers four and five."[28]

The importance attached to having children, respect for older people and the observance of memorial rites almost certainly preceded the relatively recent introduction of Islam among the Kazaks. Among other peoples of the Caucasus and Central Asia these are probably pre-Islamic values too. It is an open question whether such values and their related practices derive from cultures with a pagan/shamanistic worldview or from cultures influenced by Nestorian Christianity or even Judaism.[29]

I suspect that these or similar values were already contained within these cultures but their external forms were shaped by the later introductions of Islam, Buddhism or Christianity. What is important is not the outward expression but the inner meaning of these

Even the smallest villages usually have their memorials to the war dead: this is the one in Lombovozh

customs within any particular culture. In some instances there have been superficial influences from other religious traditions: I have heard of examples of borrowings from Christianity among some Buryats, from Islam among some Kalmyks and from Buddhism among some Kazaks living adjacent to Kalmyks.

However, the question is not one of the postulated 'origin' of a particular custom but of its meaning now for the participants concerned. For example, some nominally Muslim Tatars or Buddhist Buryats have begun to celebrate Christmas by putting up a Christmas tree and by the giving of presents — but the event seems to be perceived more as a kind of children's holiday than as a Christian festival. To a large extent, the adoption of this apparently 'Christian' custom was facilitated by the Soviet practice of putting up a 'New Year' tree with a red star at the top. Existing values within a culture not only influence the adoption or otherwise of customs from other ethnic groups but also the meanings attached to those customs if they are adopted from outside.

Among many peoples of Russia and Central Asia — almost irrespective of whether they are nominally Muslim, Buddhist, 'pagan' or Christian — there is a widespread and deeply-rooted sentiment that the dead should be honoured and remembered with appropriate memorial rituals: therefore it would not be surprising if such pervasive feelings should have found expression also within Soviet Communism. To what extent did such attitudes influence the almost ubiquitous establishment of war memorials in virtually every village and of 'shrines' with 'eternal flames' to honour the 'glorious dead' in every town?[30] Those who sacrificed their own lives so that others might live are honoured and remembered in a quasi-religious manner — compar-

A Kalmyk bride: often newlyweds will also pay their respects at the local war memorial

able to the Christians' gratitude for Christ's sacrifice in dying so that others might have an opportunity to receive a 'new life'.

Those who visit the war memorials do so with a mixture of feelings but these often stem from a sense that it is important to remember those who had died. It is not uncommon for those visiting the war memorial to focus their thoughts particularly on their own departed ancestors, especially those who had suffered in the war, and to describe their feelings at the memorial in terms such as "it pinches my heart." After a wedding often a newlywed couple will also visit the war memorial and place flowers there. The desire to remember the past is reflected in many ways — such as attention to anniversaries, the importance attributed to the celebration of birthdays and an emphasis on some aspects of history cultivated by the Soviet education system. Memorialism is also cultivated within most homes by photograph albums. Often individuals within the family have their own albums showing scenes from that person's life. Very commonly the albums are shown to guests: partly this provides a topic of conversation, but some guests expect to be shown the photo album and ask to look at it. A sense of nostalgia for the 'Good Old Days' of the Soviet past is nowadays not uncommon among some of the senior generation but many younger people have such feelings too; for instance, on her thirty-sixth birthday a Russian woman spontaneously decided to get out not only her photo albums but also many other mementos of her past, such as certificates and badges awarded to her at school and in the Komsomol.

Considering the Communists' preservation of Lenin's embalmed corpse on public display, with queues of people lining up to pay homage, how short a step is it from this to something which might be labelled as 'ancestor worship'? Not dissimilar to a respect for the dead is the emulation of heroes from the past.[31] Epic poetry commemorating heroic deeds of the past, such as the Kyrgyz Manas or the Kalmyk Zhangar epic, have become inextricably linked with the worldview and culture of such peoples. In the 20th century, heroes of the Soviet Union and the 'personality cults' surrounding Lenin, Stalin and some more recent leaders are apparently following in similar traditions. The transformation of ancestors into gods among peoples like the Mansi has perhaps had some modern 'secular' parallels within an atheistic state.

There is a Russian saying: "If you don't know your past you have no future"; this seems to apply not only individually but also to some extent to the social group as a whole. Similarly among contemporary Kazaks, "it is considered as a personal responsibility to remember where one comes from . . . Tribal identification reflects just . . . the region where one's ancestors come from."[32] Another kind of social group memory was that kept alive by the Crimean Tatars after their deportation to Central Asia, so that the older generation's experiences of suffering would be meaningful also for younger generations brought up outside of the Crimea.[33]

All peoples have ancestors and usually feel a need to preserve their memory in some manner. However, in a multi-ethnic state in which intermarriage between members of different ethnic groups is not uncommon, a question arises: "Who are my ancestors?" — and, by extension: "Who am I?" How one chooses to remember one's forebears is partly an expression of one's own ethnic identity. It also has important implications concerning a wide variety of other culturally appropriate forms of religious behaviour. This issue will be the focus of our attention in chapter five.

5
Who am I?

"Even though you are my only daughter, if you ever do that again, I'll kill you myself!", shouted Galiya's father. "Now get down on your knees before Victor and beg him for mercy."

Sobbing her heart out, Galiya fell on her knees in front of her husband and begged him to forgive her. He had discovered that his wife had been having an affair with another man. She had slipped out late one night after Victor had come home drunk and had fallen into what Galiya had thought was a heavy sleep. Seizing her opportunity to see her boyfriend, she had phoned him and had slipped out of their flat. To her dismay, Victor had woken up in the early hours of the morning and discovered his wife missing. When she returned wearing a rather provocative style of clothing he guessed what had been happening.

As a teenager, Galiya had been a keen member of the Komsomol and had no interest in boys. She wanted to serve the Party. However, at university she met and fell in love with Victor, her first boyfriend, whom she married. Her Tatar parents, especially her father, had been unhappy at her marrying a Russian, but gradually they began to accept Victor. Ironically, Galiya's adultery brought her father and husband closer together.

At first everything seemed to be going well. Galiya felt fulfilled in her marriage, and the birth of their son made their happiness together complete. Victor insisted that their son have a Russian name, to which Galiya agreed because the boy was, after all, half Russian. However, this was merely the tip of the iceberg. Whenever Galiya wanted to watch a Tatar television programme or to listen to a radio broadcast in Tatar, Victor insisted that she turn it off as soon as he entered the home. Galiya began to tune in to the Tatar channels in secret. Victor did not allow her to speak Tatar in his presence or to have any items around the home which were obviously connected with the Tatar culture. He would not go out with her to the Tatar theatre, which Galiya loved, even though they provided a Russian translation through headphones.

Gradually their lives drifted further apart. Both of them had their own jobs and increasingly separate social circles, and Victor's work often took him away from home for extended periods in other cities. Galiya's loneliness was compounded by a deep sense of rejection; she felt that her value as a person was undermined by the fact that her own husband would not accept her for who she was and that he was embarrassed by her Tatar ethnic identity. He could not change her Tatar name but his rejection of every other aspect of her Tatar identity compounded her sense of isolation and need to find acceptance and love from another source — in this case, from a Tatar man.

In a case like this, tensions between husband and wife reflected not only their ethnic identities but also a whole set of values and expectations. In any workable marriage the partners need to agree on at least some basic values such as their attitudes towards family planning and the rearing of children. Choices include the languages used in talking with the children and the social customs which are taught to them. Cultural values also influence attitudes towards family planning: for instance, in Turkmenistan there is a great emphasis placed on the value of having a son, so that some families with daughters will continue to have children until a son is born. (This is one of many cultural influences behind the greater population growth rates among some peoples of Central Asia and the Caucasus as compared with peoples of European Russia: it is not necessarily a 'religious' issue, even though there is a strong Islamic influence among the Central Asian peoples, in so far as population growth among the Tatars and Bashkorts is similar to that of the Russians.)[1]

Galiya and Victor had one child, but another had died in infancy so they had been reluctant to try for a third one. However, their surviving son was aware of the ethnic tensions between his parents. He could understand Galiya when she spoke in Tatar but he never used it actively and always spoke in Russian. Despite being half Tatar himself, he was growing up with a feeling that somehow the Tatar culture was second-best and perhaps even something of which he was ashamed.

Ethnic identity and feelings of self-respect

"Previously . . . I reckoned myself as a Tatar woman, and I wasn't ashamed of that," commented a Tatar woman from the city of Nizhnyekamsk. "As a future Muslim woman, I kept myself within the boundaries of the *shariat*, as far as I had been taught. Now, whenever there arises a special need to know the Tatar language, I am boundlessly grateful to the Lord that he did not allow me to forget my native language. Living among Russians, he didn't let me become Russified — that is, to forget my own culture, language and traditions. Praise God!

"Often I hear from [other Tatars] that formerly they had been strongly embarrassed by speaking in Tatar and by the fact that they belonged to the Tatar nationality. Of course, they have begun to forget their native language and from childhood have been conversing only in Russian. On the one hand, such activities are fully understandable. If a child doesn't know Russian from childhood then he doesn't have a future — he won't be able to study in school, and then at an institute, and so on. And there's another particularity too: if from an early age he hasn't learned to speak in Russian, then naturally he will have a strongly expressed accent in his pronunciation. This only means that people around will laugh at him, but psychologically to carry this burden is very difficult not only for a child but also for an adult."2

Lilya's comments highlight the dilemma faced by the son of Galiya and Victor. He is growing up with a sense of embarrassment regarding his mother's ethnic identity and the fact that he himself is half Tatar. None of us is responsible for our own parentage but it is more difficult to accept our heritage if it is regarded as a stigma in the eyes of other people. Whether the feelings of stigma come from being born out of wedlock or being born as a Tatar, there can sometimes arise feelings of shame generated by the attitudes of others. Even though this is a 'false guilt' (or 'false shame'), because the child is not personally responsible for the actions or ethnic identity of his or her parents, the feelings of shame and the social consequences arising from them can persist throughout the person's lifetime and continue to influence forms of behaviour.

A Tatar woman reading Islamic literature at a mosque in Eastern Tatarstan

To some extent feelings of rejection or of being under-valued on account of one's ethnicity among some people are compensated for by a sense of pride in the positive values of their own ethnic group. Self-perception plays an important part in the way in which people select certain values as not only characteristic of that ethnic group but also as being more highly developed among them than among the Russians. Whether or not these alleged differences are 'objectively' true is not as important as the subjective belief that in these respects the values of the minority people are better than those of the Russians.

Among the Volga Tatars, for instance, in cities of both Eastern and Western Tatarstan (i.e. Almetevsk and Kazan) Tatar friends of mine have spontaneously commented that Tatars value cleanliness and have claimed that Tatar villages are kept more clean and tidy than Russian villages. They allege that it is noticeable in the way that rubbish is not left lying around, and also in the general appearance of the houses, windows and yards. Similarly, a Tatar woman in Kazan says that she often (about once a week) sweeps clean the corridor outside the front door of her apartment even though six other families live along the same corridor and she has no obligation to be the one to do it. Nevertheless, she feels that her social environment, including the corridor outside her front door, has to be neat and tidy, so she does it anyway.

Who am I?

Such perceived differences — if they are indeed genuine contrasts — are statistical rather than absolute differences because of course there are always exceptions. An instance of this comes from this same Tatar woman who says how she avoids dropping litter in the street even if there is no litter bin available, and she considers this also to be a sign of her Tatar background: however, just a few days previously I had been with another Tatar woman who apparently took it for granted that there was no problem in dropping rubbish on the street — and did do so. Another value sometimes regarded as characteristic of the Tatars is a Tatar work ethic. Some Tatars compare themselves with Jews in terms of their aptitude for hard work combined with a certain shrewdness; they joke that Jews fear whenever a Tatar is born.

A third Tatar value is respect for their parents, particularly in terms of a very high regard for their mothers. This is especially strong among the generation now in their fifties who were born during or shortly after the Second World War, when virtually every family had lost at least one male member, often more, and some mothers were left with the burden of bringing up perhaps four or five children single-handed. However, in the general Tatar culture there is also a strong feeling of love and respect towards mothers, as reflected in popular Tatar songs: one informant estimated that about fifty per cent of Tatar songs are in praise of one's mother.[3]

On the other hand, traditionally the father was to be feared. Children wanting to make a request of their father often went through the mother as an intermediary. The father had a high social status in the family and had to be accorded respect; on buses in Kazan one can sometimes see an elderly, retired couple where the husband is sitting on a seat while his wife stands next to him — thereby showing her respect for her husband. This is not so common among younger generations, but even in families where the father might become drunk or abusive, there is nevertheless often a feeling that others in the family should speak of him, at least in public, with respect and in a positive fashion. One could ask whether perhaps this cultural value placed on respect for the father mixed with fear of the father might influence Tatar perceptions towards God as a being who is to be feared but also accorded respect.

Respect for parents is the most conspicuous of the Tatar values related to the family, in so far as other values are less clearly distinguished from those of the Russians. One aspect of this among Tatars is the apparent lack of a strong preference for having children of one sex rather than the other — although there is a saying that if a family has first a son and then a daughter then this is a gift from God. This contrasts with the cultural preference for male offspring in, for instance, Turkmenistan.[4]

As far as one can tell from impressions, in the absence of statistics known to me, divorce rates in the cities of Tatarstan appear to show little difference between Tatars and Russians. In Tatar villages, however, there is still quite a strong negative attitude towards divorce and especially towards divorced women. One Tatar woman suggested that this difference between rural and urban areas is probably a reflection of the fact that the farming economy in itself places demands on women to fulfil expected duties and to work within the family unit. In such families the husband is seen as the foundation stone around which the family economy is built, and there are little or no alternative sources of employment for woman outside of the domestic economy. By contrast, in the cities there are many more sources of female employment available to divorced women, so it is easier for divorcees to live separately from their former husbands. This difference in available alternative forms of employment possibly contributes to, or else is reinforced by, the reduced social stigma attached to divorced Tatars, and in particular divorced Tatar women, in the urban environment.

However, several informants agree that in general Tatar families tend to be more closely attached to each other than Russian families. Certainly my impression from visiting a number of Tatar homes is that it is not uncommon for Tatar relatives to be present in the home and to visit one another relatively frequently. An observation of this kind is perhaps to some extent explicable by the fact that in Tatarstan itself — where most of my Tatar friends live — the Tatars are almost by definition closer geographically to their places of birth and to their relatives than are at least some of the Russians who have moved into the area from elsewhere. However, among the Tatars friends also visit relatively often. This can also be a pattern among some Russian families, and among both Russians and Tatars can be compounded by feelings of mutual help among family members whereby, for example, they will make sure that they visit and help in whatever way they can an elderly member of the family who is living off a very meagre pension or a younger woman whose husband had been killed unexpectedly in an accident and is struggling to bring up her young children. These kinds of economic factors affect both Tatars and Russians, but there also seems to be a relatively greater degree of co-operation and mutual help among Tatar families than among many Russians. This is of course a very subjective impression which would be very difficult to substantiate by concrete statistical evidence, which as far as I know is not at present available. Nevertheless, the fact that there seems to be a qualitative difference between Tatars and Russians in terms of more closely-knit family bonds among Tatars is attested by several different informants.

Self-perceptions of ethnic identity in terms of values are likely to become increasingly important in the 21st century because these kinds of cultural values are deeper and probably more meaningful as indices of ethnicity than some of the more 'superficial' symbols of ethnicity (such as styles of dress or traditional foods) which are increasingly becoming out-dated. Neither is it meaningful to try to impose linguistic or religious criteria, as definitions of ethnic identity, on younger generations of Tatars who in many cases no longer have a knowledge of the language and religion of their ancestors. However, many of the younger generation do still subscribe to the kinds of cultural values described above, and can recognise them as conveying something

Who am I?

meaningful in terms of what it means to be a 'Tatar'. Moreover, these kinds of values are also recognised by both Muslim and Christian ('baptised') Tatars: therefore perceptions of ethnicity based on shared values can overcome problems inherent in the attitude that to be a Tatar means one has to be a Muslim. The same kinds of issues are also relevant to many other ethnic groups, whether they are traditionally Muslim, Buddhist or pagan.

Some people find difficulty in articulating their self-perceptions and their cultural values but they nevertheless recognise such attributes if they are pointed out. This often seems to be the case among some of the Finno-Ugric peoples, who have sometimes been afraid to express in public their feelings of being treated as second-class citizens.[5] Several Udmurts, for instance, have told me that they have sometimes been teased by Russians for speaking in Udmurt or have been called 'Votyak' (the old term for Udmurts) as a term of abuse. Rural Udmurts have felt embarrassed at speaking their own native language when visiting Izhevsk, the capital of Udmurtia, where Russians account for 69.4 per cent of the population.[6] A recent psychological survey indicates that 23 per cent of Udmurts (as compared to only four per cent among Russians and Tatars) feel their nationality is humiliated or belittled by other ethnic groups.

An Udmurt woman brings her piglets to sell at the market: in the winter they have to be kept warm in the back of the car

Moreover, 28 per cent of Udmurts (compared with nine per cent of Russians and one per cent of Tatars) felt that often representatives of their own ethnic group do not esteem or respect themselves.[7]

Among the Khanty, Mansi and other peoples of the north, the 'Internat' boarding school system has done a great deal of harm to the self-esteem of the indigenous peoples. Children of herdsmen and fishermen were taken from their families during long periods of their formative years and brought up in Internat boarding schools where they were cut off from their own cultural and linguistic heritage. They were even forbidden from speaking in their own languages! Even if their native languages were taught as an academic subject for some classes, they were forbidden from using it in real life in ordinary situations. Is it not surprising that they should begin to feel as if they were second-class citizens?

Traditional cultural values provided a sense of personal dignity and self-esteem — whether through being a good wife and mother or else through having an ability to provide for one's family by hunting and fishing. However, the erosion of their traditional ways of life and the incorporation of their social systems into a wider economic and political framework meant that the older value systems were often broken down too. Unemployment and relative poverty began to reinforce any sense of rejection and sometimes produced feelings of embarrassment about one's ethnic identity. Perhaps it is not surprising that some should have tried to find an 'anaesthetic' in vodka. Nevertheless, there still remain among the Khanty and Mansi several positive cultural values (such as a respect for nature, a contentment with a simple lifestyle and the respect accorded to older people) which can potentially help to rebuild their sense of national dignity and self-worth.[8]

A Mansi grandmother — respect for older people remains a Mansi value which has not yet been eroded

The maintenance of ethnic and religious boundaries

Feelings of being 'second-class citizens' are more likely to develop in situations where the ethnic group is not only in a minority but is also regarded negatively by the majority. Those who feel both threatened and also relatively powerless often adopt 'avoidance' strategies to reduce situations of conflict or stigmatisation. Strategies of this kind are more commonly adopted by those from smaller ethnic groups whose access to power and influence is generally fairly limited as compared with more numerous ethnic groups.

Sometimes people have sought to conceal their ethnic identity altogether, especially in situations of persecution. In the past this strategy was sometimes chosen by Jews who had fled from the anti-Jewish pogroms in Russia or from the persecution in Nazi Germany. To a considerable extent there are still anti-Semitic elements within the Russian population, as expressed most notably

Who am I?

Gypsies begging on the streets of St Petersburg — stigmatised ethnic groups and marginalised religious groups both fear persecution from Russian ultra-nationalists

through the rhetoric of Vladimir Zhirinovsky. Considering the real danger of further persecution breaking out if such ultra-nationalistic political parties were to come to power, it is not surprising if some Jews try to pass themselves off as Russians and to reduce the extent to which their Jewish identity is obvious to others.

Some religious groups have responded in a similar manner, using defensive strategies to protect themselves. However, whereas ethnicity is more or less fixed from birth (except that there is still some choice for those of mixed backgrounds), religion often involves greater personal choice. Even so, many religious groups behave like ethnic groups in their strategies for survival in the face of opposition.

Some religious and even ethnic groups have had to go 'underground' through force of circumstances. During the Communist period official recognition was given to certain ethnic groups and to a number of religious organisations but others were not recognised.[9] For instance, the 34,665 Nenets people live mainly in the tundra region of north-west Siberia, in the Yamalo-Nenets Autonomous okrug (district) which includes the Yamal Peninsula. During the Communist period the officially recognised language of the Nenets was that of the reindeer herders in the tundra regions. However, an estimated 5,000 or so Nenets belong to the Forest Nenets, who live in the forested areas further south, mainly in the Khanty-Mansi autonomous okrug. They too traditionally lived by reindeer breeding, hunting, gathering and fishing, but their language is different from that of the tundra Nenets. Forest Nenets was not an officially recognised language during the Soviet period, and neither were they recognised as being a distinct ethnic group in their own right. Therefore they had no assistance from the State in maintaining or developing their language. As far as I know, the only literature which has been produced in Forest Nenets is a *samizdat* ('self-printed') news sheet produced by a poet named Yuri Vella. In developing this, Yuri had to introduce two new letters to their alphabet in order to represent sounds used among the forest Nenets but lacking in the standard orthography of the tundra Nenets. Some other ethnic

A 'Mountain Jew' from the Caucasus

A Nenet man

groups were previously recognised but in later census data were counted as members of a similar ethnic group: for instance, the 1926 census recorded 10,034 Beserman living in northern Udmurtia and the Slobodskoy region of the Kirov oblast but in later census data they were counted as Udmurts. Their language is a form of Udmurt but also contains a number of Turkic features; it is thought that probably their ancestors had been a group of southern Udmurts who had received a strong Turkic influence from the Bulgar state and perhaps the Chuvash people but who later migrated to the Cheptsa valley of northern Udmurtia.[10] Even though they adopted Russian Orthodoxy in the 18th century they have preserved alongside it various traditional ('pagan') folk beliefs and aspects of ancestral veneration. Among some Beserman, Islamic influence has remained to some degree in customs such as inviting a mullah to officiate at funerals and certain other events.[11] In recent years there have been attempts to revive the distinctive Beserman culture and to remove any feelings of stigmatisation which had previously been associated with the Beserman ethnic identity.

In a similar manner, some religious groups had previously not been officially recognised by the authorities and had to struggle in order to maintain their identity. Often they faced persecution of one kind or another: though this was particularly the case during the Soviet period, there had also been during the Tsarist period instances of intolerance towards certain religious groups. As a result of such persecution the 'Old Believers' had in effect become like ethnic communities in their own right, often practising marriage within their own group and preserving many specific practices which distinguished them from outsiders: such symbols of religious identity became virtually equivalent to 'badges' of membership and similar to the symbols (such as dress and ornament) which are often used to distinguish particular ethnic groups.

The 'Old Believers' regarded themselves as preservers of the true form of Orthodoxy, and viewed as heretical the officially approved form of Orthodoxy. History repeated itself in the 20th century when another group of Orthodox Christians maintained that they were preserving the legitimate Orthodox faith and rejected the way in which leaders of the official Orthodox church made compromises with the Soviet authorities. These Orthodox 'dissidents' formed an 'underground' movement known as the True Orthodox Church which was severely persecuted both by the Soviet authorities and by the officially recognised Orthodox church.[12] Some True Orthodox preachers referred to the Communist system in terms such as "satanic" or the "Bolshevik devil," and their publications as instruments of "great deception."[13] Similarly, in Yekaterinburg I have been shown 'Old Believer' works of art dating from the 17th century which depict Peter I as the Antichrist.[14]

Nina Bykova, a Chuvash Baptist Christian who was imprisoned from 1966 to 1969 on account of her religious activities

Religious groups of this kind strive to preserve doctrinal purity but have the danger of becoming isolated from society at large. Some Russian Protestant groups have manifested the same tendencies towards exclusivism and isolationism. During the Communist period, and to some extent still today, the 'unregistered' Baptists and Pentecostals have tended to isolate themselves from the State, fearing the persecution which might arise again in the future if they were to register themselves and thereby accept the restrictions which could come from government legislation. Behaviourally they are not dissimilar to the 'Old Believers' or the True Orthodox even if theologically they are at the other end of the doctrinal spectrum within Christianity.

There seems to be a tendency within Russian religiosity for a certain degree of dogmatism to develop. This tends to focus on outward symbols more than on inner spirituality. Ordinary Russians often see religion more in terms of aesthetics — such as the artwork of the Orthodox churches — rather than as a system of belief or as a framework for ethics. Its truth value, or even the truths which are represented by the icons, are not felt to be as important as the actual icons or paintings themselves. Of course those with a more discerning spiritual awareness do recognise the importance of doctrinal truths and ethical values, and also value the experiential side to religion in terms of a personal commitment to the faith and a desire to speak with God and to hear his voice.

Nevertheless, the ordinary person's perception of religion in terms of aesthetics, and of religious experience as mediated through the beauty of art or music, is also characteristic of attitudes among those more religious people who tend to focus more on outer form than on an inner relationship with the Almighty. For instance, in many Orthodox churches there are likely to be elderly women who will make sure that female visitors put on a headscarf or men take off their hats. Very similar attitudes can also be found in some Protestant churches: a Baptist man in Chuvashia, for example, questioned the fact that I was wearing a gold wedding ring because his group regarded such ornamentation among Christians as being "worldly."

Whatever the outsider might think about such attitudes, the maintenance of such symbols actually serves an important function within the group in terms of demarcating boundaries between 'insiders' and 'outsiders'. Forms of dress and ornamentation, songs and certain types of (religious) language all serve to reinforce the distinctiveness of a particular group — whether that group is a religious or an ethnic one. There are also more subtle behavioural codes within each community, whereby it forms its own sub-culture: examples in some religious groups include deference towards elders or the leadership roles of men. Unfortunately, challenges to these symbols of the sub-culture by those who view them as outdated or inappropriate, can be regarded by leaders of the religious group as being forms of rebellion against divinely ordained rules. Therefore it is not surprising that there is a tendency for fission within certain religious groups (whether 'Old Believers' or Baptists) over points of 'doctrine' which, in practice, are often also symbols of the sub-cultures of these particular groups.

Smaller communities of this kind serve important social functions in modern cities where there is a considerable amount of distrust of other people and also a great deal of loneliness. Distrust fosters isolation, but religious groups promoting shared values of honesty and virtue stand out as havens in which people feel able to trust one another. In large impersonal cities the religious organisations are similar to the smaller face-to-face interaction of village life. It can be argued that for most of human history mankind has lived in smaller social groups — extended families, bands, villages and so on — and that mankind is naturally more inclined towards such a form of social interaction than the more anonymous and lonely types of social intercourse found in modern cities. People often try to find smaller groups in which they feel accepted and valued for themselves. Such an acceptance is often found within religious communities, some of which even use the language of 'extended families' (that is, fictive kinship terms like 'brother' or 'sister') to promote feelings of belonging to one another.[15]

Intermarriage and ethnic or religious identity

A preference for endogamy — marriage within the group — is not uncommon among certain ethnic and religious groups, and in particular those which have felt themselves to be under threat from outsiders. Christian groups, including 'Old Believers', Baptists and Pentecostals, strongly encourage marriage with others sharing their faith, justifying this on the grounds that the couple needs to be compatible spiritually as well as in temperament and other ways.[16] Unofficial statistics among Crimean Tatar communities indicate that at least 90 per cent of marriages are within their ethnic group — higher than the proportion among most other Central Asian nationalities.[17] This promotes not merely family cohesiveness but also a strong sense of identity as an ethnic group. The development of such a high rate of endogamy is understandable in the light of their history of deportation en masse from the Crimea to Central Asia, Kazakhstan, Siberia and the Urals. On 17th-18th May 1944 they had been rounded up into cattle trucks, packed into locked waggons under military guard, with thirst and suffocation contributing to many deaths on the way. An estimated 200,000 to 250,000 Crimean Tatars were deported on the pretext that they were 'traitors' who had helped the Germans during the war. Therefore they received little or no sympathy from the local population in the areas to which they were deported. Rather than having done anything to help those fighting against the Soviet Union, the Crimean Tatars (and other deported peoples from the Northern Caucasus) were victims of Stalin's fear lest such peoples should potentially give assistance to an invading army.[18]

The influences of 'tradition', 'religion' and parental approval are among those which have been invoked to explain the relatively high rate of endogamy within one's ethnic group which is reported among the Kyrgyz, Kazaks, Azerbaijanis, Uzbeks, Turkmen and Tajiks.[19] When marriage does occur between members of different nationalities, often it is between Muslim ethnic groups which had been less clearly distinguished from each other prior to the Soviet 'divide and rule' approach.[20]

The Volga Tatars (who are distinct from the Crimean Tatars even though both speak Turkic languages) constitute a possible exception to this pattern, partly because four centuries of close contact between the Russians and the Volga Tatars have led to many cases of intermarriage between them. This is acknowledged by a Russian proverb which says: "Scratch a Russian and underneath you'll find a Tatar." Over the centuries a certain amount of intermingling between the Europeans and Asiatic peoples of this region has

Azerbaijani men from Georgia

resulted in the Volga Tatars having a wide range of outward appearances, from those with fair skin and blue eyes to those with more Central Asiatic features. On the basis of outward appearance it is difficult or impossible to distinguish most Tatars from Russians; often their dress, types of accommodation, furnishings and other external characteristics are also indistinguishable. Therefore a limited repertoire of symbols are still available for use as cultural markers to distinguish Tatars from Russians. Predominant among these are language and religion.

Even these, however, are not necessarily sufficient in themselves, because many younger Tatars, especially in the cities, speak Russian better than Tatar and, as a legacy of Communist atheism, do not regard themselves as having any particular religion. For at least some Tatars with whom I have discussed such issues, it seems that what basically defines their Tatar identity is the fact that they have 'Tatar' written as their nationality in their passports.[21]

In Tatarstan itself, Russians are the majority ethnic group in several major cities, such as Kazan, Zelenodolsk, Bugulma, Chistopol and Yelabuga.[22] Tatars often constitute the locally predominant ethnic group in smaller towns and district (raion) centres, where in the 1960s-70s women had been more likely than men to reject intermarriage.[23] Nowadays, however, in the cities of Tatarstan, at least among those mixed marriages known to me, it appears to be more common for a Russian man to marry a Tatar woman than vice-versa. These impressions are borne out by study conducted in Ufa, where the Tatars — who constitute about 27 per cent of the city's population — have the highest proportion of those who prefer marriage to a partner of the same nationality. However, the percentage is higher among the men than among the women, because 60 per cent of Tatar men but only 50 per cent of Tatar women expressed a preference for a Tatar spouse.[24] However, in interpreting these statistics it is important to bear in mind that Ufa is the capital of Bashkortostan (Bashkiria), where the Bashkorts constitute a further 11.3 per cent of the city's population. In view of the relatively close cultural and linguistic links between Tatars and Bashkorts, there are some local residents who tend to minimise the distinctions between these two groups. I do not know if this attitude is more pronounced among women, but it was certainly typified by a group of women whom I met in Ufa. When I asked them if they were Bashkorts or Tatars, they replied, "We're all the same, all mixed up together," and they refused to categorise themselves as one or the other ethnic group.

A Tatar man and his Bashkort wife with their daughter: this nominally 'Muslim' family also put up a 'Christmas tree' in their home (but call it a 'New Year tree'!) and have observed the 'Santa Claus' tradition

In Kazan, the capital of Tatarstan, about one in three marriages are of mixed nationality.[25] Many others in the city who regard themselves as either Tatars or Russians also have grandparents or other relatives of a different nationality. However, even if the offspring are officially either Tatars or Russians, they are aware that in themselves they are of mixed descent.

Nowadays it is possible to choose which nationality would be adopted by the children of mixed marriages. Often the choice has been governed by the nationality of the father, whose surname would also be adopted.[26] However, during the Communist period (and to a certain extent still today) it was often thought to be more advantageous for children of mixed marriages to choose to be Russians.[27] Sometimes a Russian nationality has been adopted by offspring of marriages in which neither parent was Russian but the child was brought up speaking Russian as the language of the home. To some extent the trend towards 'Russianisation' is now being reversed, partly encouraged by 'positive' discrimination in favour of ethnic minorities: among the indigenous peoples of the Russian north and far east there are economic incentives to retain a non-Russian identity not only because of 'traditional' rights, such as

After Atheism

Examples of inter-ethnic marriages: (Clockwise from top right) A Lak man from the Caucasus with his Russian wife; a Ukrainian man and his Khanty wife (wearing Khanty national costume); a Tatar man and his Chuvash wife; an Armenian man with his Kalmyk wife

hunting or fishing privileges denied to Russians, but also on account of access to special funds for the peoples of the north. For instance, a family in northwest Siberia were eligible to obtain an extra grant to help with the construction of their large new house because the wife speaks Khanty. Her husband is Ukrainian but in this case the wife's nationality was a definite advantage. I have been told that even people with only one grandparent belonging to one of the indigenous nationalities have begun to register themselves as being from that ethnic minority, rather than as Russian, in order to gain access to such benefits.

To some extent Communism helped to reduce religious conflicts within families of mixed nationalities, in so far as atheism became the official ideology — one might like to say 'religion' — of many people. With the current religious freedoms, children of mixed marriages now have a choice to make regarding which, if any, religion they might want to adopt too. If the parents subscribe to a religion at all, even if only nominally, do the children follow the Islamic faith of the Tatar parent or the Orthodox Christianity of the Russian parent? An example of such a dilemma is that which was faced by a young woman named Dina.

As a child, Dina was torn between two religions. Her Tatar grandmother secretly taught her Islamic prayers but her Russian grandmother secretly had her baptised. Each warned Dina never to tell her other grandmother. They each told her that the other grandmother would kill her for not following the same religion as that grandmother.

Throughout her childhood and teens Dina pretended to her Tatar grandmother that she was a Muslim and to her Russian grandmother that she was a Christian. Finally, her inner conflicts were resolved when she heard about the Baha'i faith — a religion which respected both Christianity and Islam. When Dina became a Baha'i, she found a faith which encouraged her to respect both Jesus and Mohammed and to study both Christian and Islamic scriptures as well as Baha'i writings. Her conflict between ethnic identity and religious faith seemed to have been resolved.[28]

The revival of religious and ethnic identity

Since the break-up of the Soviet Union it appears that to some extent there has been a reversal of the previous tendency for children of mixed marriages to prefer to adopt a Russian nationality if one of their parents was Russian. In Kazakstan there are the beginnings of a movement to encourage Kazaks to marry other Kazaks rather than Russians. It appears as if this desire to promote their own ethnic identity and to reduce intermarriage has been fuelled by the break-up of the former USSR and the emergence of a new sense of national pride as members of a fully independent state. While this process is most conspicuous among members of the newly independent nations, there are similar processes at work elsewhere. To a considerable degree, the Soviet system itself was responsible for the creation of this sense of national identity through a policy of 'divide and rule' which entailed the sometimes rather artificial fabrication of separate national histories and of distinct literary languages or alphabets. In the Northern Caucasus, the Adygei, Cherkess and Kabardins were originally one ethnic group, known as Circassians in English, speaking an indigenous Caucasian language. However, they have not only been split up into their present-day 'nationalities' but also isolated from one another even further by the administrative ploy of making the Cherkess share a region with the Karachai and the Kabardins with the Balkars — both the Karachai and the Balkars speaking Turkic languages unrelated to that of the Cherkess (Adygei) or Kabardins. In the same way, the Turkic people of the northern Caucasus were divided into the Kumyk, Karachai, Balkar and Nogai people, who nowadays regard themselves as distinct ethnic groups.[29] Similar processes appear to have taken place among the Turkic peoples of Central Asia, who had previously been less differentiated but now tend to regard the rather artificial borders of the Soviet period as definite national boundaries between distinct peoples.[30]

A Mari man in traditional Mari costume at a 'pagan' ritual. Both the costume and the ritual are vehicles for the expression of Mari ethnic identity

In the Volga-Ural region the Bashkorts were divided off from the Tatars, and probably the same process occurred in the separation of the Komi-Permyaks from the Komi: there are ten different dialects among the Komi and four among the Komi-Permyaks but I am informed that many Komi Permyaks regard themselves as basically Komi.[31] The Buryats of southern Siberia were divided into three distinct administrative units but some would now like to bring these together again.[32]

However, for each of these peoples any sense of national consciousness needs symbols to express and represent that identity. Among many of these peoples ethnic identity is expressed largely through the use of their national language, the preservation and revival of traditional customs and the expression of religious practices. The use of religion as a vehicle to express national identity is particularly important in two ways: firstly, as a rejection of the atheism imposed by the Communists, and, secondly, as a form of ethnic self-expression which is different from that of the Russians.

Who am I?

To some extent this even applies to ethnic groups who had previously been largely Christianised through Russian Orthodoxy, such as the Udmurt, Mordvinian, Chuvash, Mari and Komi peoples of European Russia. Among the Mari in particular, and also to some degree among the Udmurts and Chuvash, there are still remnants of pagan religious practices. In each of these areas there are groups of nationalistically-minded people who see the revival of paganism as a form of promoting their ethnic identity. There are similar movements elsewhere, for instance among the Abkhaz people of Georgia who had also preserved many aspects of pagan practice. Whereas the Abkhaz want to distinguish themselves from the Georgians, in the Russian Federation the desire is to promote expressions of national identity distinct from that of the Russians.

A Chuvash couple in Moscow whose way of life contains few outwardly visible expressions of their ethnic identity

One of the Udmurts involved in promoting paganism commented to me that paganism is more 'environmentally friendly' than Christianity. Such arguments equate the Russians with Christianity and then criticise actions by the Communists or by the Russian state as if they were performed by Christians.

The greater freedom of religion brought about by the collapse of the Soviet system has meant that each ethnic group has begun to search again for its own distinct roots. Religion is merely one aspect of that search, but an important one because it also is capable of providing an alternative reason for living and a basis for morality and interpersonal relations. In the next chapter I shall present a few case studies of some of the ways in which members of various ethnic groups have sought to find religious expressions of their national roots.

6
Finding one's roots

Islam

It is Friday lunchtime. At a small mosque in Yekaterinburg forty-two men gather for prayer. In a separate house behind the mosque, about a dozen women pray separately but through a loudspeaker on the wall of the room where they meet they can hear what is going on in the main mosque.

The mosque has been open for eight years. Previously, the devout Muslims had held their services at the Muslim cemetery outside the city. It had entailed a two hours' bus journey and then a walk on foot. During the winter months they had performed their prayers in sub-zero temperatures, outside in the snow, even though most of those present were elderly people.

The Yekaterinburg mosque is situated to the south-west of the city centre, in an area which has at times been inhabited by gypsies. There used to be a bus which went near to the mosque but the route has been discontinued, so the believers now have to walk three or four kilometres on foot. The mullah is not sure whether or not the discontinuation of the bus route had anything at all to do with the presence of the mosque, but he has written to the city authorities asking for a bus service to be re-established, at least on Fridays.

Mullah Hazrat had been brought up in a devout Muslim family. His grandfather had been a lecturer in Islamic culture who possessed a large library with many copies of the Qur'an. As a child, Hazrat had studied at home for five years instead of going to the state school, until the city authorities visited his father and forbade the tuition of religious topics. Hazrat continued to study Islam by himself from older people and from some of the staff of the Spiritual Directorate for Muslims in the Russian Federation.[1]

The Mullah says that on the whole he now has no serious problems with the State officials in Yekaterinburg. Previously he had helped to build a mosque in Chelyabinsk, where the city and district authorities had been very co-operative. They had even helped by providing some of the building materials.

This mosque in Yekaterinburg caters for the nominally Muslim Tatar and Bashkort population, but only a tiny proportion of them actually attend this mosque. Yekaterinburg's combined Tatar and Bashkort population amounts to about 65,000 — about five per cent of the city's population.

Mosque attendance is also relatively low in Kazan, the capital of Tatarstan, where Tatars account for 40.5 per cent of the city's population. Farid Heidar

Finding one's roots

Mullah Hazrat and Tatar men praying in the mosque on a Friday lunchtime

al-Salmani, who is in charge of religious education at the Marjani mosque in Kazan, claimed in 1991 that during the month of Ramadan about 30,000 to 35,000 Muslims attend one of the city's functioning mosques. He said that for other special festivals attendance is about 15,000 to 16,000, while on a usual Friday about 5,000 to 6,000 people attend mosque. However, independent observers consider these figures to be exaggerated: on an average Friday lunchtime only about 300 men and 200 women attend the city's main mosque. For the important festival of Kurban Bayram — known in Arabic as the feast of 'Id — over a thousand Tatars did attend an open-air service, but this was certainly exceptional.

Since 1988 the Marjani mosque has been organising classes on Islam for teenagers and children. The sexes are kept separated, but the total number of boys studying at two different mosques is no more than about one hundred. To a certain extent Islam is being used as a means for cultural revival and

After Atheism

reaffirmation of national identity among at least some Tatars.[2] However, after some seventy years of atheistic propaganda, most Tatars have only a 'nominal' allegiance to Islam and most of them know very little about Islamic belief or practice.

The headquarters of the Spiritual Directorate for the Muslim peoples of European Russia and Siberia is located in the Bashkort city of Ufa. In 1992 I attempted to interview the Mufti but was told that he was unavailable; instead I spoke with Rashit Gilmanov, the head of the 'international department' of the Muslim spiritual board, who told me that over the previous two to three years there had been a rapid increase in the number of mosques, the total having risen to over 700 as compared with a 1980 figure of only 94 mosques for the whole of the European USSR and Siberia. Almost every day the Mufti had to attend the opening of a new mosque somewhere in the region, but I later learned that many of these are actually small local prayer houses — buildings converted from other uses — rather than being purpose-built mosques. A few purpose-built mosques are being built in the region, usually with the help of funds from the Middle East, although inflation and other factors have meant that sometimes they have been unable to complete the half-finished structures.

Mosques in Tatarstan: in Kazan

The mosques are financed, he said, mainly by ordinary Muslims and by the sale of Qur'ans and religious literature. There has been some private but not official (i.e. governmental) financial support from abroad: for example, on 10th January 1992 the Central Islamic bank in Jeddah, Saudi Arabia, gave 1.5 million US dollars towards the construction of mosques. Gilmanov said that a delegation from Jeddah would come that July to attend the opening of a new mosque in Naberezhnye Chelny, coinciding with the claimed 1,000th anniversary of the introduction of Islam to that area. New mosques were also being built in cities such as Samara, Omsk and Irkutsk, while elsewhere old

Finding one's roots

buildings are being bought and converted into mosques. Muslim private schools were being organised too in Kazan and other cities. However, towards the end of 1994 the construction of about 200 mosques was halted on account of insufficient funds.[3] Many of them remained as empty shells but in 1996 it was reported that the number of newly built mosques in Tatarstan was growing daily.[4]

Gilmanov's estimate of mosque attendance in 1992 was that in Ufa visitors to the mosque next to the headquarters of the spiritual directorate on a normal working day number "about 100 who pray." However, when we then visited that mosque we discovered about a dozen people, most of whom were Uzbeks passing through Ufa (especially while on pilgrimage). Only two local (Bashkort or Tatar) men were there, receiving instruction from an imam sent from Turkey to teach Islam.

Mosques in Tatarstan: in Almetevsk

According to Gilmanov, attendance at the mosque on Fridays rises to about 500 men and probably an equal number of women. However, on 6th April 1992 the mosque and surrounding streets were filled by perhaps 5,000 people who had come for the festival of Kurban Bayram. Perhaps one incentive to attend was the fact that it was also being televised and was being attended by the Chief Minister of the Presidium of the Supreme Parliament of Bashkortostan accompanied by several other dignitaries. Apparently this was the first television programme about Islam in modern times which also showed officials taking part in the prayers. However, Rashit Gilmanov was somewhat sceptical about the emergence of Islamic cultural centres and of Islamic parties because, he claimed, they represent "only the official face of Islam and not real Islam: these people do not go to the mosque and are not real Muslims."

In Ufa the prayers are said only in Arabic but the sermon is given in Tatar or Bashkort by imams of those nationalities. Five religious schools had already been set up — in Ufa, Naberezhnye Chelny, Kazan, Orenburg and

After Atheism

A Muslim teacher from Turkey [right] with one of his Tatar 'disciples' at the mosque in Ufa adjacent to the Spiritual Directorate's headquarters

Oktyabrskiy.[5] Of the 70 students at the school in Ufa, five had been to study the previous year in Turkey but 25 wanted to do so in the current year. 'Guests' from Turkey sometimes read the Qur'an in the mosques: thirty imams had already come to Ufa, Naberezhnye Chelny and Kazan. A few local people had been granted scholarships to study Islam in Algeria, Morocco and Jordan, and two of them were now teaching in Ufa. Many people attend the free evening classes held at the mosques. Some also study Arabic. Copies of the Qur'an are now available but according to Gilmanov those which had been printed in Kazan contained 273 mistakes. Others have been donated from Saudi Arabia and elsewhere.

However, even during the Soviet period there was also the presence of 'unofficial' or so-called 'parallel' Islam.[6] It was manifested most prominently in Sufi orders, especially in the northern Caucasus and parts of Central Asia. Although traditionally a male form of religion, there were indications that some female Sufi groups had developed too. One attraction of Sufism was its 'aura of forbidden fruit' — if the authorities denounced it, then it was presumably important! In 1975 it was estimated that "more than half the believers" in the Chechen-Ingush republic were adepts of Sufi brotherhoods.[7] Moreover, Islam was felt to be almost inseparably linked to many aspects of ordinary daily life, especially in rural areas, to such an extent that ethnic identity and an affiliation with Islam were felt to be virtually synonymous.[8]

Finding one's roots

Nevertheless, nowadays to some extent the nominally Muslim peoples of the Caucasus and Central Asia have begun to look back beyond their Islamic past to their original cultural heritage. For example, the Tatars, whose conversion to Islam occurred partly between the 10th and 14th centuries, are now re-examining the roots of their ancestry in the Bulgar culture which flourished in the middle Volga region in the ninth century AD. Some are beginning to reject the ethnonym 'Tatar' and to claim "we are all Bulgars!" — this slogan was painted in large letters as graffiti in the central area of Kazan. It might be significant that even before the break-up of the Soviet Union there had been a strong Tatar interest in their distant past — including the early Bulgars and the period of the Kazan khanate — as if people were trying to discover their cultural and ethnic roots.[9] The Bashkorts are also looking to their past. Some are turning to Islam but some mullahs see a threat to Islam in the form of a reverence for the ancient high god Tengri. In parts of Bashkortostan nature veneration has also survived in the worship of the Mountain Spirit Tau-Eyakhy and, to a lesser extent, the water spirit Khlu-Inyakhy and the house spirit Ui-Iyakhy.[10]

In the absence of written documentation, any form of archaeological or linguistic evidence is a clue to a people's past.[11] For instance, in Tajikistan there is an interest in their nation's pre-Islamic past, in the periods when their country was included within Bactria, Parthia or Sogdiana. As Zoroastrianism had at that time been the main religion of their people, an interest in Zoroastrianism has developed today among some younger Tajiks. This was reported to me by a Tajik anthropologist in St Petersburg named Rakhmat Rakhimov, who told me of a discussion which he had overheard in Samarkand. It was about the title of a newspaper which they were going to publish. One suggestion was to name it '*Sogd*', after the Sogdians, but one old man, who was a believer in Islam, was adamant that such a name was totally unacceptable. After a long discussion they decided to name it '*Ovozi Samarkand*' ('Voice of Samarkand').[12]

Older Tajiks have more of an interest in Islam and regard the Sogdians and other pre-Islamic peoples as 'infidels'. Nevertheless the Tajik cultural association in Moscow, which consists mainly of students and younger people, was named Sogdiana, despite objections to this name among some older Tajiks in Central Asia. When Tajik students in Leningrad (before it was renamed St Petersburg) organised their own association, one proposal was to name it 'Bokhtar' (Bactria). However, the final decision was to call the society 'Tajikistan' because nowadays nobody in Tajikistan uses the terms 'Bokhtar' or 'Sogd' as designations of their own nationality, as all call themselves Tajiks.

Tajiks are also interested in the evidence about their own roots unearthed by Tajik archaeologists. They are very interested in symbols which are actually Tajik ones, so some Tajiks have begun to imitate Zoroastrian imagery of certain kinds of birds and horses. Perhaps Zoroastrianism has more novelty value than some other religions, insofar as both shamanism and Nestorian

Christianity could also lay claims to being pre-Islamic religions of Central Asia.

At the same time, Islam is also enjoying a resurgence among many Tajiks. In many villages of Tajikistan, new mosques are being built and old ones repaired. Often the aim is to have a mosque in each *mashallah* ('parish' or 'area'). Many people have been giving their voluntary labour to help with the projects, or else providing donations of money or building materials.

Some of the younger Tajiks also regard Islam as a factor which conserves their culture — including traditional poetry, literature, architecture and sciences. For younger people, Islam represents the roots of their culture, whereas older people are more interested in the sphere of religion. Even for the older people, practice often seems more important than 'belief', in so far as they seem more concerned with the ceremonial aspects of Islam than with doctrines. Nevertheless, there is some interest in both aspects of Islam, and the doctrinal side has also been promoted in the series about Islam broadcast by Teheran radio. The fact that the Tajik language is close to Farsi (i.e. Persian) means that they are open to influence from Iran in both the social and the political dimensions of Islam. Other influences come from the Tajiks of Afghanistan, although some of those brought up in the former USSR under the Communist system are less sympathetic to the more 'fundamentalist' forms of Islam even if they view aspects of Islamic art and literature as part of their own cultural roots.

Just as some younger Tajiks are looking back to the pre-Islamic (Zoroastrian and Bactrian) roots of their culture, so a similar phenomenon has been occurring among some of the Ajar people of Georgia, whose language and culture is essentially Georgian but who had become Muslims during the three centuries when their territory had been part of the Ottoman Empire. During that period, however, the Ajars maintained their Georgian linguistic and cultural heritage and "had not interiorized deeply the Moslem religion and never were too eager to support it."[13] This gives some credence to an unconfirmed report that about five thousand nominally 'Muslim' young Ajar people were baptised together when the head of the Georgian Orthodox church visited the Ajar region in the Spring of 1989 in order to open a new church in Batumi. Apparently he was met by this large crowd of Ajars who wanted to become Christians — they were all baptised together.[14]

On the other hand, there are also Ajars who have reacted against Georgian nationalism of a kind which has refused also to give recognition to the desires of ethnic minorities within Georgia such as the Abkhaz and South Ossetians. Since the late 1980s some Ajars have therefore begun to wear veils, practise polygyny and to display other public signs of being Muslims, as well as trying to promote other links with Turkey.[15] Therefore religion can be used among the Ajars either as an expression of nationalism to assert ethnic distinctiveness as different from the Georgians or else as a symbol of identification with the Georgians in terms of their original cultural roots.

Throughout Central Asia and the Caucasus there still remain vestiges of attitudes which had been fostered by the Soviet educational system. For example, a Turkmen woman, brought up only a few miles from the border with Iran, told me how she had been educated in Russian schools where she had been taught to regard her culture as "backward" and as "like that of the Australian aborigines." However, she now realises that there are many things to value and preserve in her culture, so she is making great efforts to find out more about the traditional way of life of her own Turkmen people.

Traditionally nomadic or semi-nomadic peoples like the Turkmen and Kazaks are now relatively urbanised or else belong to collective farms so

Turkmen women

some have begun to feel that they have lost their 'roots' and no longer have a distinctive culture of their own. There are discussions about the extent they might wish to return to pre-Soviet customs such as polygyny. Some unmarried Kazak women who fear lest they might never marry are in favour of a law which would allow men to have second wives, whereas many married women are opposed to it because they do not want to share their husbands with other women.

There are also local folk customs which have their roots in the pre-Islamic religion of the Kazaks. Many of these ideas relate to fertility and illness, and may involve shamanistic rituals: some shamans were operating in Kazakstan during the Soviet period.[16] Nowadays they are more likely to be known as faith healers or psychics.

A Kazak man playing a traditional musical instrument

Buddhism

Tibetan Buddhism is the form of Buddhism which had spread among the Khalkha Mongolians, Buryats, Tuvinians and Kalmyks. Buryat Lamaism belongs to the so-called 'Yellow Hat' sect of Mahayana Buddhism, which spread to Buryatia in the 18th and 19th centuries. Before the Russian Revolution there had been 36 monasteries and 16,000 lamas in Buryatia, but by the end of the 1930s all the monasteries had been closed down and the lamas dispersed. In 1946 two monasteries were re-opened (the Ivolga *datsan* in the Buryat ASSR and the Aga *datsan* in the Aga Buryat National Okrug, with no more than about one hundred lamas) but a further ten were opened between 1990 and 1992.[17]

Few Buryats were able to visit the open monasteries during the Soviet period because they were in relatively isolated locations. Both in Buryatia and Mongolia some visitors came in order to pray for recovery from illness or to remember the dead, as described already in chapter four. Apart from annual rituals on the occasion of major Buddhist festivals, many of the visitors attended for certain life-cycle rituals, including those occurring every twelve years at the beginning of a new cycle of the Chinese 'animal years' calendar.

Finding one's roots

A Mongolian 'oboo' (cairn) — one of the motivations behind adding a stone to it is as a sign of respect to the deity of that locality

Some illnesses might be attributed to the neglect of such rituals, as reported by a 41-year-old Khalkha Mongolian woman: "Every twelve years is a difficult year but my mother didn't pay attention to it. However, when she was 61 years old that year was difficult for her: she said that she was dying in such a year because she hadn't paid attention to the tradition so before she died she told me to go to the temple in such years. She was born and died in the year of the horse. Previously this was the commonest ritual but we're only just now recovering these traditions."

Both in Buryatia and Mongolia there has been a revival of lamaism and the rebuilding of several monasteries but in the Mongolian republic still no more than a third of the population regard themselves as Buddhist 'believers'.[18] At a popular level, there are also many religious practices connected with shamanism.[19] Some modern shamans are also lamas, and nowadays there often tends to be co-operation rather than antagonism between lamaism and shamanism, even though at times in the past there had been attempts by both Buddhists and Communists to suppress shamanism. Shamans take part in rituals for particular clans or lineages, and also in ceremonies at sacred sites marked by a heap of stones called an *oboo*. (Some *oboos* also became associated with lamaism.) Shamans either inherited the power from a parent or else were taught shamanic techniques by another shaman. Nowadays, however, those who claim to be shamans no longer wear the traditional costumes and often they themselves admit that they cannot achieve a genuine trance state.

A new monastery has now been built in Kyzyl, the capital of the Tuva region. Shamanism and various kinds of divination continue to be the main kinds of popular religion. Examples of shamanic activity include animal sacrifices, worship of the spirits of the mountains and purification of the water at sacred springs. Tuvinians often fear the attacks of spirits, who are thought to bring illness or misfortune. Young children are thought to be especially vulnerable to the attacks of vindictive spirits, because their souls are said to be less firmly attached to their

Finding one's roots

bodies. Until the age of three or four, some boys are dressed as girls or girls as boys in order to confuse and deceive malevolent spirits.[20]

Throughout the Soviet period, there had been no Buddhist monastery in Kalmykia, but religious festivals and fasts had been observed in secret by Kalmyk Buddhists. Then in July 1989 a provisional temple was opened on the ground floor of a block of flats in Elista, the Kalmyk capital. Now a new purpose-built temple has been opened on the southeastern outskirts of the city. Most of the Kalmyks I have interviewed told me that they had visited it at least once or twice, mainly for special festivals. However, its location means that it is less easily accessible by public transportation. In Elista itself a statue of the Buddha provides a convenient meeting place near the city centre and Buddhist styles of architecture have begun to appear on the roofs of

Photos: Facing page: (top) inside the Buddhist temple in Elista; (bottom) a Kalmyk family in Elista. This page: A Kalmyk woman places some money in the temple offertory; (bottom) A Kalmyk man next to the prayer wheels outside the temple

some shops or kiosks but these appear to be largely an expression of nationalism rather than of religious feelings. Although many Kalmyks would tend to regard themselves as Buddhist if they considered that they have a religion at all, some feel that they have been alienated from Buddhism through their lack of contact with it during the Soviet period. The deportation to Siberia reinforced this alienation: for instance, one Kalmyk woman told me how she was confused about religious questions because she had been brought up with influences from atheism, Russian Orthodoxy and Buddhism.

Judaism

Under the Soviet system of national 'homelands' for each nationality, the Jews were allocated an official 'home' territory, called Birobidzhan, near to Khabarovsk in southern Siberia, but less than 15,000 Jews — and now, with recent emigration, possibly as few as only 3,000 — actually live there. Instead, most Jews are scattered throughout the major population centres of Russia. Nowadays a revival of Jewish culture has been taking place not only in Moscow or St Petersburg but even in far-flung cities such as Barnaul. In many cities Jewish children have been re-learning their culture through having extra-curricular Hebrew lessons or Israeli dance classes. Most Jews speak Russian as their first language and do not speak either Hebrew or Yiddish, but they recognise that to some extent their ethnic identity is closely tied up with their own language and religion, even though practising Jews account for only about a tenth of Russia's 500,000 Jews.[21] Although a religious component is by no means absent from Jewish social or cultural societies, it seems as if the focus is more on learning Israeli dances and basic Hebrew as expressions of their culture than on the study of the Torah or attendance at a synagogue (if there is one).

For example, there are about 6,000 Jews in the city of Kazan (which has a population of 1.2 million) on the Volga river, but those who have been attending the local Jewish club or actively involved in Jewish activities numbered only about four hundred.[22] A former synagogue is now used by the Tatar Ministry of Education, who seem unwilling to hand it back to the Jewish community. The stigma which had been attached to the word 'Jew' is still a major reason why many of the Jews in Kazan do not participate in Jewish festivals or attend the concerts and lectures held approximately once a month at the Jewish club which had been formed in Kazan in 1989. Those who do attend tend to be Jews who want to go to Israel or America.

One of the founders of this Jewish club was Leonid Sons, who works in the Tatar National Theatre. The other was Salomon Zeldovich, an engineer at a scientific institute, whose father, Academician Zeldovich, was highly respected in the Jewish community because he was a famous Soviet scientist. For many years Salomon had dreamed of emigrating to Israel but he had been kept in Kazan by the KGB during the Brezhnev and Andropov periods. When Gorbachev came, however, Salomon fought for his rights to leave the country.

Finding one's roots

He was granted an audience with the president of the Tatarstan branch of the KGB, who advised Salomon to "be wise" because he knew the name of Zeldovich very well. Later he was allowed to leave for America. From America he wrote that he "lives well," but the only job he could find was as a baker's assistant. The Jews of Kazan found it amusing and sad that a famous and clever engineer should end up with such a low-status job in the West. Similar reports from Israel sometimes tone down the attractiveness of emigration as an option for others in the city.

Without Zeldovich, the Jewish club in Kazan has been affected by at least three different 'enemies'. The first is fear, the second is apathy and the third is the desire for help from Israel. Israel sent many books to the Kazan Jewish community, but some of the club's ordinary members claim that the ten leaders of the Jewish club sold the books and for about two years lived off the proceeds without working. Many of the Jewish youth have now lost interest in the club and go elsewhere for their social activities. One Jew from Kazan who does have an active involvement in Jewish affairs, even though he now lives mainly in Moscow, is Yevgeny Aronson. He works for the Jewish Children's Foundation in Moscow, which, among other things, has helped Chernobyl victims by organising summer camps for children. At first the camps were for Jewish children, to teach them something of their national culture, history and religion, but later the camps were opened also to children of other nationalities, including Russians and Ukrainians. The foundation also runs a Sunday school for Jewish children, which is held near the Sokol metro station in Moscow.

A Jewish man at the front entrance to a synagogue in Moscow

Although primarily concerned for the welfare of children and the teaching of Hebrew, the foundation also helps Russian Jews to emigrate to Israel and does what it can to help them start a new life there. They do not receive aid from the Israeli government although they do have some links with a group of ten Israeli organisations who have a variety of different political leanings.

Some finance for the foundation's charitable activities comes from the sale of books which they have helped to publish, including phrasebooks and a Hebrew-Russian dictionary. They also advertise on a shortwave band of Radio

A meeting of the 'Messianic' synagogue in Moscow

Moscow and on a programme called 'Shalom' which is broadcast from Israel. In response to these appeals they have received donations not only from Jewish citizens within the CIS but even from the Council of Evangelical Christians and Baptists.

In Moscow there is also a synagogue (here meaning 'community', rather than a building) of 'Messianic Jews' — that is, Jews who believe in Jesus as their Messiah. Their meetings are led by Jews wearing traditional Jewish religious dress. Jewish symbols such as the Star of David and the Israeli flag are prominent in front of the group playing guitars and other instruments who lead the worship, consisting primarily of contemporary songs set to traditional Jewish styles of music and often accompanied by a group of dancers expressing their worship through an Israeli style of dance.

In Kazan, Israeli-style praise songs have also been commonly used by a charismatic church which has been openly supportive of the Jewish community and has been praying for Jewish people. In September 1994 this 'Cornerstone' church organised a church conference in which a Jewish musical team led worship in Hebrew and the church took a stand against anti-semitism in Russia. However, shortly afterwards, on 9th October 1994, a gang of five youths raided the pastor's apartment and demanded one million roubles (then worth about $500). At the time of the raid the pastor, Sergei Borisenkov, and his wife Julia were out at a church service but Sergei's sister, Lyudmila, was baby-sitting for the Borisenkovs' children — five-year-old Ilya and three-year-old Masha. Lyudmila said she did not have such money but the gang tied her up, said they wanted to kill her, and attempted to throttle her. She managed to survive the ordeal, but was left with scratches on her face which the police confirmed were criminal injuries. On breaking into the flat, the gang had claimed they were from Zhirinovsky's party, that they hated the Cornerstone church in Kazan and wanted it to be annihilated. They stole all the audio equipment and cassettes — together worth about $1,000 — and with a pair of scissors cut the electricity wires, which continued to emit sparks. The children witnessed all this and were left crying and frightened.

The Borisenkovs' version of the police reaction was that they received the reply: "If it's to do with politics, don't come to us, go to the mafia! Who is your protection?" When Sergei replied: "God is our protection", the policeman said he was crazy. A member of this Cornerstone church then made some discreet inquiries among people in the mafia but was told that those who broke in were probably not from a mafia organisation but instead were merely common criminals; apparently the mafia would not deal with 'commoners' — only with other mafia.

Then, on 19th October, the names of the culprits were revealed by Dennis Derevyanko, a former member of the Cornerstone church who is now a member of the LDPR (Zhirinovsky's party). Their leader, Valera, was about twenty years old and the others were all about eighteen years old; one of them was known to be mentally ill. They returned all the goods but said that Dennis Derevyanko had put them up to it — a fact which Dennis then admitted. Apparently he had thought the Christians would not go to the police and that he could get away with the raid without being punished. The raid was on the private initiative of Dennis Derevyanko and had no official link with the LDPR, which afterwards denounced their member for the robbery.[23]

Paganism in European Russia

Aspects of pre-Christian paganism have survived among several ethnic groups, including the Abkhaz of Georgia and some of the Finno-Ugric peoples of European Russia, where there had been a continuing tradition of paganism.[24] Among the Mari people of the Volga region it has served as a vehicle for the preservation of national identity in maintaining their distinctiveness from the

Muslim Tatars or the Orthodox Russians.[25] Mari paganism includes rites directed towards the spirits of the forest and fields (who are believed to inhabit sacred rocks, trees and rivers) and offerings to the spirits of ancestors who are thought to wander around at certain times of the year. To a lesser extent aspects of pre-Christian paganism have also survived among the Udmurts (in rural more than urban areas). Paganism is even less obvious among some of the rural Chuvash people, although sometimes in the popular consciousness Orthodox saints had replaced pagan deities. Only slight vestiges seem to have remained among the Mordvin.[26]

On 12th July 1998 I attended a Mari pagan ceremony in the village of Shor-unzha — its official Russian name, but known in Mari as Uncha — in the south of the Mari-El republic, near to the border with Tatarstan. I was accompanying a group travelling there under the auspices of the Mari-El Ministry of Culture so everyone had to conduct themselves in a manner deemed to be respectful. As it would be inappropriate to use the forest around the sacred place as a toilet, it was necessary to stop the minibus several kilometres from the site so that those who needed to relieve themselves could do so. Women also had to put on headscarves. On arrival at the site, visitors were expected to place a monetary donation in an offering box and then to wash their hands as a symbol of ritual cleanness before entering the sacred grove itself.

Cooking the meat in a cauldron at a Mari pagan ritual: on branches of the tree in the background can be seen clothing placed there by those praying to the gods

We were met by Alexander Ivanovich Tanigin, a '*kart*' (the term for a Mari pagan priest) who asked who the visitors were and why they had come. When a Mari professor of archaeology explained that I was conducting anthropological research, the *kart* replied that I was welcome to attend but only if I wrote "kind words" about their festival. Tanigin then pronounced a curse over me, stating that "something bad" would happen if I wrote anything

Finding one's roots

"bad" about it. Moreover, I was not allowed to take any photographs without the permission of a man who was designated to keep an eye on me. I replied: "I understand."

This man took me to one side and said that I was not allowed to photograph fresh meat, only cooked meat, and that I was not to come too intrusively close to the priests during the prayers themselves but had to stay at a distance. Later he explained that once a Russian journalist had photographed a pagan ritual elsewhere and had come up too close to the priest in a disrespectful manner. He had then written an article in a sensationalist and offensive manner which was published by a Moscow magazine. Six months later the journalist had died.

Mari women attending the ritual bring as offerings pancakes and other food they have prepared

Apparently for similar reasons Alexander Tanigin did not want me to take his photograph, whereas another *kart* named Alexei Izergovich Yakimov did allow himself to be photographed by a number of people. Yakimov said that his grandfather had also been a *kart* and that from childhood Alexei had expected to become a *kart* himself. His family had kept up the pagan rituals and so other Mari people had asked them to perform ceremonies on their behalf. Alexei commented that a *kart* "has to be clean" — an attitude which is consistent with the idea that the sacred site has to be kept pure and undefiled too.

Alexander Tanigin had compared himself with the shamans of the north who, he said, were "higher than us" because some of them could put a knife through themselves and then carry on as normal whereas the Mari *karts* could only receive revelations from the spirits through dreams. Perhaps a further influence on Tanigin's reluctance to be photographed was his belief that the shamans also did not allow themselves to be photographed.

Some probable links between Mari political aspirations and the revival of paganism as an expression of Mari national identity are indicated by the fact that the man whom Alexander Tanigin had delegated to supervise my conduct was an active member of a new Mari political organisation named 'Kugeze Mlande'. He had brought with him several copies of a recently published newspaper with articles in both Mari and Russian about the organisation and its aims. Kugeze Mlande wants not only to revive the Mari culture but also to form in the Volga-Ural region a new State confederation which would be independent of Russia. I was told that under the previous president of the Mari-El republic, Kugeze Mlande had been forbidden access to the mass media to inform the general public about their political aspirations whereas

The president of the Mari-El republic lights a candle in front of the tree where prayers are directed to St Peter

the new president, Vacheslav Kislitsyn, is taking a more neutral stance, neither prohibiting nor encouraging their activities.

As the current president was also expected to attend in his official capacity this pagan rite which I observed in July 1998, no ceremonies were performed until he had arrived. In the meantime people from local villages brought offerings of food which they presented at one or another of the five sacred places. Each site was dedicated to a different deity but they were in such close proximity to each other that to walk from one to the next took only a minute or two.

Early that morning, prior to our arrival, the animals had been sacrificed at each of the sacred places. For instance, two rams and a goose had been sacrificed at the place for prayers to the god of the ground. By the time I arrived, at each site the meat was being cooked in one or more cauldrons

Finding one's roots

suspended over fires. Around the sacred trees had been tied 'belts' on which there were traces of the blood from the offerings and into which branches had been inserted. Other branches had been laid on the ground in front of the trees in the area where local people were placing their food offerings. Visitors were kept from treading on any of these branches. On branches up to twenty feet or more above the ground were shirts, dresses or jackets which appeared to have been placed there on an earlier occasion — I was told that these were offerings by those who had petitioned the gods for healing from illnesses.

Local people whom I interviewed told me that such festivals are held twice

The Mari 'karts' (pagan priests) lead the prayers while the people kneel

a year, on the 22nd of May and "in the middle of July" (or more specifically the 12th) but one *kart* said that a third rite was held on the 19th of December. When I asked some villagers about their attitudes to the festival, they replied that they saw the rites as "part of local tradition": having grown up accustomed to performing these rituals, they said they had no particular opinions about them. People from several villages attended the rites. Those from Shor-unzha itself, which contains about 350 households, reckoned that the rites were being attended by perhaps a quarter of the village population, "or maybe more." Some local residents had contributed a goose or another animal for a sacrifice whilst others had brought along home-baked bread or other food they had prepared. Those unable to bring food contribute money.

They also explained that the date of this festival was fixed as St Peter's Day according to the Orthodox calendar. One of the five sites is actually dedicated to St Peter. The rites at each site are under the supervision of a different *kart*, one of whom told me that the prayers are intended to include people of all religious backgrounds, whether Orthodox or pagan. The village has no church so they come instead to these sacred groves in the nearby forest for a "general religious festival," where "whoever wants to give an offering is welcome to do so." Many people had brought along candles similar to those used in Orthodox churches: these were lit in front of the trees along with the food offerings.

Apart from St Peter, there are also groves dedicated to the god of the earth, to the god of lightning, to "the mother of God" and to the god "Mikol" — described as a "great god" for whom there are festivals both in winter and

Cutting up the food offerings after the Mari pagan prayer ceremony

summer. The *kart* said that the name is derived from that of St Nicholas. I received different replies when I asked if the shrine for the 'Mother of God' was supposed to be for Mary, the mother of Jesus, as this is a term for Mary often used in Russian Orthodox circles. Whereas at least one of the villagers thought the grove was dedicated to Mary, the *kart* said that this is not Mary but a "women's god" to whom childless people pray for fertility — that the deity would bestow a child. He said that they pray to the god of lightning to bestow rain at times of drought, and to the god of earth for general health and happiness for all, including one's relatives and others.

This Morkinskiy district in the south of the Mari-El republic is reported to be one of the most traditional Mari districts. However, I was told by this *kart* that the rite was being attended also by representatives from each of the other districts of the republic. They come three times a year to offer their greetings

Finding one's roots

out of "respect." It was only when the president himself arrived that the main rite could begin.

On his arrival, the president was welcomed by the *karts*, who offered him and his associates a portion of some of the pancakes which had been brought as offerings. He and his colleagues took some and ate it. Then he was escorted to the grove dedicated to St Peter where he lit a candle in front of a tree. Meanwhile all the local people had gathered around in a semicircle on their knees ready for the prayers. They are predominantly women at the front but some men are also in attendance towards the rear of the crowd. All the *karts* are male and they remain standing while the laity kneel.

Alexei Izergovich Yakimov, one of the 'karts', and a Mari woman sample some of the cooked meat

The main priest placed a deep bowl of cooked meat in front of the tree and then began to recite various prayers in the Mari language. This went on for several minutes, after which they began to cut up the pancakes and other offerings which had been laid in front of the tree alongside the lit candles. After this the food is offered around to all those present for them to eat; I had the impression that those who had presented offerings to the deities brought around their own offerings for other people to sample. As people were starting to leave and the vessels were being tidied up, a ladder was put next to one of the trees and a young man climbed up to place a new bundle of clothes on one of the branches.

Aspects of paganism have survived to a lesser extent among the Udmurts and Chuvash too. According to the criteria used, between five per cent and 30 per cent of Udmurts might be classified as 'pure pagans', between 15 per cent and 40 per cent as Orthodox and the remainder as 'syncretistic'.[27] An artist named Semen Nikolayevich Vinogradov and a number of others have set up a committee which wants to revive paganism and establish a pagan sacred shrine. One member of the committee, Albert Razin, has felt it necessary to write a defence of his advocacy of paganism in the face of opposition from at least one Udmurt member of the Orthodox clergy.[28] Even though ten out of seventy-five Orthodox clergy in Udmurtia are ethnic Udmurts, and there are also Udmurts among the active lay participants in the Orthodox church, nevertheless "there is a kind of unreflective, unwritten code towards the Udmurts in the church which prevents them from feeling at home there ... for an Udmurt to become Orthodox means psychologically to reject his national interests."[29] In some Protestant churches there are also Udmurts and the evangelical 'Philadelphia' church includes a regular weekly meeting in the Udmurt language.

Whereas many Udmurts within Udmurtia itself adopted Russian Orthodoxy alongside some aspects of paganism, the older faith survived more strongly among the Udmurts of Bashkortostan and Tatarstan.[30] Four centuries ago some Udmurts fled south and east away from the armies of Ivan the Terrible. Settling among the Bashkorts and Tatars, these Udmurts clung more strongly to their paganism because it became an important symbol of ethnic identity. Other markers of ethnicity were their styles of dress and their Finno-Ugric language, although they also began to use many Tatar loan words. Udmurts in Bashkortostan and Tatarstan also tended to encourage endogamous marriage with other Udmurts because Udmurt girls who married Tatars were normally expected to become Muslims and to bring up their children as Muslims. Recently Udmurt pagan worship services have been held in the village of Kaishnabali in Bashkortostan but have attracted opposition from Muslim activists.[31]

In a number of instances, pagan and Christian elements have become intermixed. For instance, shortly after observing Orthodox rituals at Easter, some villagers pray to the pagan gods, asking that the coming summer would be blessed with a plentiful harvest, and that everybody in the family would be happy and healthy. Later, in the autumn, the men thank the gods of nature for having granted them a good harvest.

The day chosen for these rites can vary from one village to another, but people say that nowadays the rituals tend to be less elaborate than in the past. One relatively more complex ritual has survived in the Udmurt village of Varklend Bodya in the Agryz district of Tatarstan, where, in the course of a week-long annual festival, they pray each day to a different god. On the first day they offer prayers to the pagan High God and on subsequent days focus their attention in turn on the god of the earth, an evil god and also the spirits of those who have died, differentiating between those who died locally and those whose deaths occurred elsewhere. They offer meat and gruel on a fire, as burnt offerings, believing that in the smoke the offerings would ascend to the High God. By contrast, the offerings to the god of the earth are left in a

Finding one's roots

Udmurts at a sacred grove at a secret place in the forest where pagan rituals are conducted

hole in the ground. The same village also holds a coming-of-age festival for young people aged between fifteen and seventeen, when the whole village comes together and prays for divine blessings on the young people. The festival is held on separate days for the young men and the young women. However, in Udmurtia itself this custom seems to have disappeared.[32]

Sacred groves are still used for religious rites in some Udmurt villages, especially outside of Udmurtia itself. In 1998 I was taken to one such grove, which could only be reached on foot via paths which were not very obvious but were marked to some extent by incisions on the trunks of trees. Its location was kept secret to outsiders: a Russian man who had lived in the village for eight years told me that he had learned about the shrine only that week, and then only because of the visit by some researchers (including myself). I had not asked to visit the shrine but it was arranged for me; however, out of respect for the confidence of the local Udmurts I am refraining from specifying even the name of the village. The grove was situated outside the village and well away from the nearest road but adjacent to a sacred stream.

Visitors to this grove were expected to put a silver coin of some kind at the base of the venerated tree. The bands around the tree and the offerings placed in them reminded me of those at the Mari rite described above. A few metres in front of the tree was a spot where at some time a fire had been lit, and I presume that the site was also used for animal sacrifices. One of the women who escorted us to the sacred grove also mentioned that elsewhere in the village is a place where figurines are kept, but she did not feel at liberty to take us there too.

Top: A pony is prepared to be sacrificed at a Mansi ritual. Middle left: A black shroud is put over the Christian cross on a nearby grave while the ritual takes place. Middle right: The pagan spirits are invoked by the man acting as shaman. Bottom left: The pony's throat is slit and the head is removed. Bottom right: Two chickens are also sacrificed.
Facing page — Top: The pony is skinned and carved up. Bottom: The local Mansi people gather round and prepare to eat the meat.

Finding one's roots

The recent revival of paganism among the Udmurts and Mari is to a large extent an expression of national identity. From a casual acquaintance with some of these supposedly 'pagan' Udmurts, I have the impression that paganism has now become more of a local tradition than a strongly-held belief system.  Perhaps this is partly due to the prolonged influence of atheistic Communism. After many decades of Russification, these peoples have lost many former symbols of ethnic distinctiveness. In the towns national dress is no longer worn except for special occasions and members of the younger generation often speak Russian better than Udmurt.

Now that religion is no longer prohibited, some Udmurts are looking to the pre-Christian pagan faith of their ancestors as another vehicle for expressing their ethnic identity. They want to show that they are distinct from the Russians and that they have a national heritage, identity and traditions of their own.

Paganism in Siberia: the Mansi

The little colt tied up to a tree looked frightened as the man approached. It seemed to sense what was about to happen to it, and struggled in vain to get away. The man covered the colt's back with a special cloth to dress it in preparation for the sacrifice.

Two men held the reluctant colt firmly by its halters as a third man announced in Russian: "This is our faith!" He then proceeded to chant in his native language, invoking the pagan gods of the Mansi people. The spirits were being invited to attend the sacrifice held in their honour. As each spirit was invoked, the man waved a strip of cloth over his head in a circular fashion.

Then one of the attendants took an axe and beat its blunt end down on the colt's head to stun it. As it lay unconscious on the ground, its throat was slit and the blood ran out into the earth. Three other men standing nearby who were holding chickens then decapitated the hens, the headless bodies of which continued to move around for a short while until they fell motionless.

Once the colt was dead, the men began to skin it and cut it up. At this point, one of them asked the Russians filming the event to stop taking photographs. An inference was that the Mansi were going to get drunk on vodka while eating the sacrifice and they did not want their behaviour recorded on film. The main participants in this ritual were men who had come from the reindeer-breeding settlement of Saranpaul, but the Mansi women spectators were allowed to participate afterwards in the feast.[33]

This sacrifice took place at 5am on the 26th of June, 1993 in the Pushkin park of the town of Berezovo. Near to the place where the sacrifice was held is an ornate grave — that of the daughter of a nobleman who had been exiled to Berezovo during the Tsarist period. It was noticeable that prior to the pagan ceremony a black cloth had been draped over the large cross above the grave to conceal this important sign of Christianity during the pagan rite.

Many of the 8,500 Mansi people live in the Berezovo district of north-west Siberia. Traditionally they were reindeer herders, hunters and fishermen, whose culture and religion was very similar to that of their Khanty neighbours to the east and south. Under the Communists their religion had been suppressed but now there are the beginnings of a revival of pagan worship in the area. There are also revivals of paganism in many other parts of Siberia, including the Sakha republic.[34]

However, the Mansi sacrifice described above was not conducted strictly according to traditional practices because most religious specialists — the shamans — were killed by the Communists during the 1930s. This was partly to suppress religious observances and partly to eliminate potential leaders of anti-Communist revolts. (There was indeed such an uprising which was ruthlessly suppressed and resulted in the loss of many Khanty lives.)

Nevertheless, one important religious leader survived because he was blind

Finding one's roots

at the time. Later, he passed on his knowledge to his son, Korstin Tarasavich, who now lives in the village of Sosva. Korstin describes himself as a "keeper" of the native religion — rather than a "shaman." He compares himself to the pope, saying that he has responsibility for ensuring that other shamans use their powers correctly.

Korstin told me that although his father was blind, he had unusual sensory abilities: he would "just know" the identity of a visitor without a word being said, or know when there was an obstacle in front of him while going through the forest, so that he never stumbled. He was also reputed to be able to hear an elk two kilometres away, when even a dog could not hear it. In 1947 they had a secret religious meeting when, after praying all night, Korstin's father told the hunters in the morning which way to go to find game and in which direction the wind would be blowing.

Korstin Tarasavich (right), who describes himself as a "keeper" of the Mansi native religion

Korstin remembers a time when as a child he had toothache but through a kind of hypnosis he was persuaded by his father that there was no pain. Now Korstin himself thinks that often the shamanic "powers" to reduce pain were through the power of persuasion. When he was 47 years of age, Korstin was summoned to the Regional Party Committee and asked about his father's activities. Korstin replied that the old man rarely travelled and then met only with elderly people. By that time Korstin himself had been taught only some factual information about the identities of "keepers of the tradition" but not how to operate as a shaman himself.

As a child Korstin was christened for the sake of his grandfather, who preserved his Orthodox icons after the churches were closed. At school Korstin wore a cross but couldn't resist his teacher's offer to exchange it for a tie-pin. Throughout his life Korstin was surrounded by Party functionaries, and was one himself for many years. Nevertheless, he says that his own religion persisted "at a subconscious level" despite his atheistic upbringing.

When asked about his own experiences of the supernatural realm, Korstin replied that he has no fear when out alone in the wild. No matter how early in the morning he goes out, he takes no dogs with him and might see the traces of a bear but the bear itself will have gone. He thinks that the bear also feels something about Korstin, a sense that this man would not shoot — but of course he has not yet had an encounter in which to test this out! Korstin then went on to talk about his father's prediction that the world as the Khanty knew it would come to an end in the year 2014 and that from the year 2000 women would be allowed to go to the Khanty sacred mountains and would start to dominate in the religious sphere. Then the Khanty understanding of their world would be overturned.

Again I asked Korstin about his own experiences of the spirit world. In reply he mentioned how he is unable to explain dreams of his in which he is a goose or is flying. Neither can he explain dreams about needing to catch horses but being unable to find them. However, he then remarked that he does not attach much significance to these dreams.

Korstin says that he always feels in advance if a person is for or against him, referring particularly to his struggles on behalf of the aboriginal peoples. Apart from this, he admits that he has very little power left. He has not been able to restore the soul to a dying person, and says that to do so depends on the dying person having a strong faith that the presence of the 'keeper' can bring back a lost soul. Once he was called to a dying person, but he said that the local people did not understand the process and should have cut a rope before fetching him. He said that when he arrived, the soul had already left. The Khanty believe that a man has five souls and a woman four. If four of a man's souls have left the body but the fifth is still present — diagnosed by placing a mirror against the lips — it might be possible for a clan member to keep the fifth soul in the body and restore life.

Korstin would now like to make the sacred place of the bear clan at Vizhakary, on the Ob river, into a sacred place for all Khanty and Mansi peoples — whose languages and religions are very closely related.

In Khanty-Mansiysk, the regional 'capital', other Khanty and Mansi people are proposing to turn Vizhakary into a tourist site, with new buildings and facilities for tourists. They would like it to become a kind of folklore museum where they might hold folk dances. It is claimed that such dances would have little or no 'real' religious meaning.

A third viewpoint was expressed to me by Konstantin, the Khanty young man who believes a curse has come on his family through their allowing outsiders to visit their sacred place in the forest. He feels strongly that sacred places should not be disturbed unnecessarily, because to do so would also disturb the local spirits. In his opinion, any kind of interference at all with Vizhakary, whether as a folk museum or as a national shrine, would be intrusions into what it is now — a place visited only by a few believers. In support of his position, he claims that since 1989, when Korstin had started on his project of trying to make Vizhakary into a cultic centre, something very unpleasant had happened to almost all of Korstin's male relatives. Many of them had died or fallen ill, and all had suffered in some degree.

Korstin had indicated to me that a man named Pyotr was another "keeper

Finding one's roots

of the tradition," but when I met Pyotr he gave no indication of this in response to my leading questions about his religious experiences.[35] What 70-year-old Pyotr did say was that he himself had been baptised at an Orthodox church in Tyumen and that the Mansi have no bad feelings towards Russian Orthodoxy. His worldview, however, is largely that of Mansi folk religion:

"God is one god," said Pyotr, "but he has many sons who live in separate places around this region. There are many names of the sons of the main god. We Mansi have a form of worship: we do not cross ourselves, but we bow towards the place where the household gods are enshrined [on a shelf behind a curtain in the corner of the living room], and we turn round three times clockwise, in the direction of the sun's course. We also have special gatherings once a year when all the old men gather and make offerings for all the people to live happily, to be healthy, and for the future to be good. We used to kill reindeer, but there are no reindeer in this village now — all the reindeer here were taken to Saranpaul [because of Khrushchev's collectivisation policies in the 1950s]. There is a special place in the forest where we used to pray at a grain store raised on legs. Now this has just started to revive.

"In 1990, on the 4th of June the first public festival was held in the village of Kimkyasui, with the help of the village council. All the people gathered there, with the women in their best dresses: it was beautiful. The festival went on for two days. We made an offering of a bull which was brought in specially. Everyone went to the place to pray for a good life. Now there's a piece of cloth left forever there, with a coin tied in it as a kind of offering.

"We also had traditional competitions for people, with prizes for running, dancing and so on. Now we want to have an annual festival. At that first festival in Kimkyasui we had no images of the gods because they were normally kept at the sacred places in the forest. What we have on our god-shelves are pieces of cloth about half a metre long which have been kept from our ancestors. We don't open it, and don't know what is kept inside. Once a year we gather the parents together and everyone brings their cloths which we put in the sacred corner of the house. Then we make our offerings and pray. As there's no longer any reindeer here, we may kill a cow or an elk. Previously the tradition was to place a reindeer before the god as an offering, and people may eat the meat or even sell it, but must never leave it to rot. In September we make an offering of a horse."[36]

Pyotr would not reveal anything more of his role as a keeper of the tradition. Later he and Vasili Vadichupov, a member of the Kimkyasui council of elders, made sure they were present when in Kimkyasui I then visited a 68-year-old woman named Anna. It seemed as if Pyotr wanted to ensure that Anna did not divulge any secrets.

At first Anna denied that she spoke much Russian, but after a while she began to speak fluently in Russian with a Russian woman who was accompanying me. Apparently she was observing a custom whereby she should not speak directly with a man. What she said about the village festival

Inside a Mansi home — the 'god-shelf' in the corner under the roof is concealed by a curtain

was generally very similar to what we had already been told by Pyotr.

Afterwards we met up with Lyudmila Ivanko, a Russian sociologist, who had been in the village 'club' talking with the local young people while I was with Pyotr and Anna. Lyudmila said that the local youth knew exactly what the village elders were doing: "They'll just drink vodka with the visitor and tell him the same old stories they tell everyone who comes." It was becoming clear that what was being presented as the 'public' face of Mansi religion was not necessarily what went on under the surface. For the older generation, however, the preservation of a distinctive ethnic identity was linked in with their concealment of traditional secrets.[37]

Similarly, in 1993 a public 'bear play' ritual was performed in the open-air museum in Khanty-Mansiysk. Veneration of the bear had previously been widespread in Siberia but had been suppressed by the Communists.[38] This 'bear play' appeared to be largely for the benefit of Russian 'tourists', but I was later informed that the 'real' sacrifice would be performed in the forest at night for those who were part of the 'inner circle'.

It appears as if the revival (and perhaps preservation) of Mansi culture and religion has been partly through the use of a veil of secrecy. What is shown to outsiders is not necessarily the true state of affairs. A false facade can be presented towards Russians and other outsiders to deflect them from enquiring more closely. What is presented to outsiders is the public festival but not the secret sacrifice, or else an 'official' account of public religion rather than details about the offerings to the spirits which are wrapped up in blankets and tied onto trees near to the Mansi graveyard. In the same way, women are forbidden from seeing figurines (household gods) stored out of sight below the roofs of Mansi homes.

Similar behavioural characteristics can also be revealed in non-religious settings — such as in Anna's pretence not to speak Russian well. Another example was recounted to me by a Russian doctor who had lived for many years in the region. He had been fishing with a local Mansi man who was very competent at fishing but whose behaviour suddenly changed when a police launch came on the scene. They were doing nothing illegal, and the police were merely performing their routine duties, but the Mansi man began to act as if he were stupid and hardly understood Russian. Once the police had gone, his behaviour reverted to that of his normal self.

On the wall of another Mansi home is a box containing cloths in which coins have been wrapped as offerings to the spirits

Therefore it seems as if the preservation of a distinctive ethnic identity has been linked in with a tendency to conceal certain traditional secrets or aspects of behaviour. In recent years there has been a revival of some public rites, such as the sacrifice I witnessed in Berezovo in 1993, which was explicitly intended to show the Russian spectators that "this is our faith." Nevertheless, a veil of secrecy remains over other religious rituals which are apparently practised more by the older generations, and of which some younger Mansi might be unaware too.

A most closely guarded secret is the location of some sacred places in the forest. Certain men have the responsibility to kill any non-Mansi outsider who even inadvertently discovers the location of such a place. I was informed of this fact by a local official after I had returned several times to the Mansi region and had been helping local people through trying to improve the water supply in one village and through always bringing with me aid in the form of medicines for the local clinic, children's clothes and toys for a local kindergarten. My friend was concerned lest I should discover the location of one of these sacred places and thereby put my life in jeopardy. He therefore spoke privately with one of the local leaders who had the responsibility to kill any outsider who finds a sacred place. This man agreed to grant me the status of an honorary Mansi.

The 'Old Believers'

Orthodox Christians who had opposed the changes introduced to the Russian Orthodox liturgy in 1653-54 by Metropolitan Nikon found themselves confronted by a decree in 1683 which made them into enemies of the state.[39] Their only hope of escaping persecution was to flee to remote parts of the Urals, Siberia and other peripheral regions of the Russian realm in order to perpetuate the older forms of worship.[40]

One of these communities had been established in the town of Nevyansk, about two and a half hours' drive north from Yekaterinburg. In 1990 I had been allowed to visit the town, but only after special permission had been obtained from the KGB because of the presence of an armaments factory (which was being converted for civilian use). Perhaps because I was the first English visitor to the town for many years, perhaps since the Revolution, I was treated like a VIP. After a tour of the local museum and a special slide show put on for my benefit, I was taken for lunch at the restaurant in the town centre. When we arrived, we found that it was specially closed for my visit! No ordinary citizens could eat there that day, only the English visitor and his companions. Specially for my benefit, a fresh bar of soap and a clean towel had been put out. Moreover, the table was nicely laid with attractive doylies, flowers and complete sets of cutlery, including knives (which are absent in many Russian restaurants). Cheese and sliced cold meat — items at the time unavailable in the shops — had "somehow" been made available for this "VIP". My Russian companions also found it amusing to see how the restaurant had been done up specially for me — and at the end of the meal they were happy to follow the suggestion of the proprietress that they could take away with them the leftover meat, cheese and bread.

After a visit to the 'leaning tower of Nevyansk' (where I had to promise not to take any photographs of the armaments factory surrounding the tower) we went to visit Isaac Mamontovich — whose name means 'Isaac Son of the Mammoth'! He was one of the leading figures among the Old Believers in Nevyansk. At the age of 86, he was truly an 'Old Believer' — in both senses of the term. He had four daughters, two sons and two sons-in-law. Isaac showed me old hymn books using the old forms of musical notation and for my benefit he sang a scale and one of the hymns.

His wooden home was in a traditional Siberian style, including a place for sleeping just above the stove. Outside, his covered courtyard was scrupulously clean: Old Believers have a reputation for regarding cleanliness as next to godliness.

In the Urals region the 'Old Believers' had often become owners of mines and factories, who employed other Old Believers to work for them. In 1939 the Soviet authorities closed down the chapel belonging to the local group of Old Believers, but they were allowed to reopen it later that year after two men

Finding one's roots

Ivan Mamontovich (left) in 1990, showing some 'Old Believer' manuscripts to the author

went to the authorities every day for two months and insisted that they be given the keys to their chapel. At that time there were 500 Old Believers in the town. The authorities then ordered them to elect a council of twenty people, and among them to select a 'revision commission', of which Isaac was the chairman. He was therefore viewed as a kind of local official because this was done at the command of the authorities. One of his tasks was to compile a list of their religious possessions. These could have been confiscated by the State, but they were allowed to continue using these books and objects while being aware that the State still had the list and might confiscate these things at any time. Such a threat meant that the local congregation tried to avoid further confrontations with the State authorities.

Nowadays, more women than men generally come to the Old Believer meetings in Nevyansk. Isaac commented that the men are more lazy, and might say, for example: "Gregory is at home, so why should I come?" In general, it is now easier than before for them to hold services, but Isaac commented that the State still has the list of their religious possessions. Isaac did not give me an estimate of the size of their average congregations, but he said that for major festivals they might have more than 100 people at a service.

Not far away from Nevyansk are the 'Merry Hills', which until 1957 were a place of pilgrimage for Old Believers because their graveyards are there. The authorities then forbade the holding of this pilgrimage but it was revived in 1990. However, not many knew about it so few came, whereas before the

Revolution about five to six thousand people used to come from all over the country, some of them on foot.

In the Old Believers' chapel they use only the old versions of scriptures and prayer books: they have no Bible at all, but have three New Testaments and twelve special books for the services appropriate to each month, plus other books for use at festivals. All these books were listed in accounts by the State's revision commission in order to estimate their price, so they are included in the official list along with any other precious items. Only these old books are used in their services. Isaac himself has an old Bible and has no need for a modern translation. He also has personal copies of a *Golden Month* book and one called the *Prologue* — a collection of the lives of saints, homilies and so on arranged as a set of readings throughout the year.

Isaac has a "nodding acquaintance" but no special contacts with the 'mainstream' Orthodox priest, whose church is just outside the town. This priest was not in Nevyansk when I visited his church, but I was told that his regular congregation consists of a small number of mainly older people. The woman in charge of the church kiosk had a few icons and an Orthodox newspaper for sale, but no scriptures. She told me that the church had no Bible at all.

'Mainstream' Russian Orthodoxy

During the late 1980s and early 90s the sudden popular 'fashion' for having an Orthodox baptism or for wearing a cross around one's neck also seemed to be a reflection of a desire among the Russians to rediscover their ethnic and religious roots. However, as an ethnic group the Russian people not only constitute the majority nationality within the Russian Federation but had also been the largest individual nationality within the former USSR (even though their proportion of the overall population had been declining on account of the higher birthrate among peoples of Central Asia and the Caucasus). Even though there is some local and individual diversity within Central Asian forms of Islam or Buddhism, the very size of the Russian population in itself means that it is capable of supporting a relatively wider spectrum of organised religious groups than tends to be the case among smaller ethnic groups. Russian Orthodoxy is often seen as an expression of Russian national identity but Orthodoxy is by no means a monolithic entity. Therefore it seems appropriate to devote the next chapter to a more detailed examination of aspects of Russian Orthodoxy.

7
The Orthodox kaleidoscope

"Over twenty years ago I had wanted to have my daughter christened in the Orthodox church but my parents had been opposed to the idea because of the influence of Communism. Now we have freedom of religion, so two years ago my daughter and her husband themselves chose to be baptised in an Orthodox church. Having already been married by a State ceremony, they are now saving up to have a church wedding too."

This account by a Russian woman living in the Urals region is typical of the attitudes of many people. They had retained a respect for the rituals of the Orthodox church but had been afraid to risk their jobs or to face questioning from the KGB if they showed an open interest in religion, let alone go through a rite such as baptism. However, in the late 1980s and early 90s the loosening of State controls on religion meant that many Russians were requesting baptism in the Orthodox church.[1]

On 16th June 1991 I attended one such service where about twenty people, young and old, both men and women, were baptised. The service was held at the Church of the Ascension in Yekaterinburg (then known as Sverdlovsk) in the Urals region. I was informed that many people had been asking for baptism so such services had become frequent occurrences. However, I was told by a committed Orthodox lady that most of these apparent 'converts' were "religiously untaught" and knew very little about the faith which they were apparently accepting.

Nevertheless, it seemed to me that such people recognised the spiritual dimension to their lives and were responding accordingly. The weakening and eventual collapse of Communism as an ideology had left a spiritual vacuum which demanded to be filled. There were unanswered questions about the meaning of life, about moral values, and about whether or not there is a life after death — and, if so, a judgement of some kind. Orthodoxy seemed to provide a sense of reassurance and meaning, even if the meaning of the words and of the mysterious rituals remained obscure or incomprehensible.

However, to some extent it was also a way of rediscovering their cultural roots. The ancient rituals, the traditional styles of architecture, the use of priestly robes and incense, the artistic styles of the icons, and perhaps even the very fact that Old Slavonic was used for the rituals all contributed towards a feeling that this was something which was authentically 'Russian'. A sense of history was important after the Communists had destroyed many of the

After Atheism

symbols of the pre-revolutionary Tsarist state which had also become symbols of the Russian culture. Orthodoxy symbolised the Russian soul — indeed, the very fabric of what had been called 'Holy Russia'. For some people Orthodoxy is important as an expression of the Russian culture, almost irrespective of whether or not they believe in the existence of God: rather than regarding it as a system of belief, such people see in Orthodoxy a vehicle by which to rediscover their cultural roots.

Some couples, who might have been married several years already, have also begun to have a second wedding in an Orthodox church and to receive Orthodox wedding certificates. Perhaps some of this stems from a genuine faith, and certainly from a kind of religiosity, but it seems that this too is also mixed in with a search for national identity and for a cultural continuity with the pre-revolutionary past.

An Orthodox church in Yekaterinburg

Similarly, in the late 1980s and early 90s it became fashionable among many Russians to begin wearing a cross around the neck. Although some wore it as a genuine mark of religious belief, many wore it simply because their friends had begun to do so, while others seemed to regard it as a kind of 'good luck

An Orthodox priest on a rare visit to the village of Berezovo is surrounded by a crowd who ask him many questions about Christianity and the afterlife

The Orthodox kaleidoscope

An Orthodox baptismal service for adults in Yekaterinburg

charm'. In such cases, the cross did not represent anything particular about their beliefs, but to some extent it was used as a symbol of one's national identity in so far as it was worn by Russians and others belonging to nominally Christian ethnic groups. (A similar expression of ethnic identity, but using a different symbol, is illustrated by a Tatar woman in Moscow who wore a crescent moon around her neck. When asked her ethnic identity, she initially said she was a Muslim and only when asked in more detail said she was Tatar. The pendant might also have been used as a charm against the evil eye but at the same time served as an expression of her ethnic identity as a Tatar in Moscow.)

Another important symbol of Russian national identity was the Tsar. These two nationalistic symbols — Church and Imperial Family — have now come together in Yekaterinburg, the city in the Urals where Tsar Nicholas II and his family were murdered. According to local informants, recent research has shown that the room of the Ipatiev house where the execution took place was actually within the boundaries of an Orthodox wooden chapel which had formerly been on that site; in other words, it is claimed that the royal family were actually killed on consecrated ground.

Since 1990 there have been plans to build a magnificent new Orthodox church on this site. Many Orthodox Christians in the city gave generously towards the project. The Moscow Patriarchate, Sverdlovsk city council and the Architects' Union of Russia together organised an international competition to select the project's architect. This position was awarded to Konstantin Yefremov, who showed me his proposed design and told me how it had been inspired by a vision:

"I'd been working on the shape of the church . . . and couldn't find the shape of the cupola, having worked out a countless number of variants . . . Then, at about 1.30am, I put out the light and prepared to sleep, but what happened then was totally outside of my volition and if it hadn't happened to me personally I wouldn't have believed it.

"I was lying on the bed under the blanket but suddenly I was somehow sitting, my legs had come under me, my head was

Konstantin Yefremov's model of his proposed church

thrown back and my face was staring towards the sky (although I was in my apartment and directly in front of my gaze must have been the ceiling of the flat, but I didn't see it). All the rest of what happened most quickly I saw not with my eyes but I felt with all my being . . . It was somehow as if I existed, but on the other hand as if somehow I didn't. Everything took place outside of time and space, although my legs were tired and swollen . . . and I understood somewhere far off deep in my consciousness that what was happening was going on for quite a long time.

"Into my head, directed towards the sky, it is as if there came . . . a pole, the diameter of which must have been a little more than that of my head, stretching endlessly long upwards and somewhere there in the endless height it suddenly sharply opened up into a . . . sphere; this sphere presented itself as a shimmering still space, visually perceived from itself — like a starry sky or snow twinkling under the moon on a winter's night. In this sphere there was neither heat nor cold, but a feeling that there these concepts are redundant: there it was a staggering purity, stillness and freedom from everything superfluous that surrounds us in our everyday worldly life. I experienced an astonishing, indescribable sensation of blessing, happiness and tranquillity. From my eyes welled up tears, but not of sadness but of joy and contentment. I began to articulate a prayer to our Lord Jesus Christ — the prayers were expressed from my subconscious, they were simply personal, not formal [literally 'canonical'], it was some kind of spontaneity, and I was fully conscious during this. It was so good for me, that I can't convey it all.

Konstantin Yefremov

"After some moments I asked: 'Lord, will the church be in this place?' And as an answer I didn't hear anything at all, but I saw at a glance, before my gaze was formed a panorama of the existing Hill of the Ascension with the opposite bank of the city lake of the river Iset, but in that spot where now stands a cross on the place where there had been the cellar of the Ipatiev house where the royal family were killed, there rose up a white church with a golden cupola. I understood that in spite of everything the church would be built on blood. I asked one more question: 'Lord, will I see the church in this place?', thinking that perhaps the church would be built after a century or two — that is, after my death. In answer I saw that very same panorama, but the church was even whiter, and the cupola sparkled even brighter. Recognising that it was obvious that I must continue the work on the church, I asked: 'Lord, reveal to me the secret of the cupola.' In that moment it seemed as if I was lifted high up above the ground and suspended in the air. To the right of my view I saw . . . a golden ball, the 'helmet' of the cupola, and around it, streaming vividly, were golden tongues of flames.[2]

"I understood that the cupola is a flame of fire and there we see in front of our gaze a hot flame of light. Suddenly I was struck by the simplicity and clarity of the answer to that on which I had been working for so long in vain.

"After this I suddenly became so light and joyful and peaceful. Again my prayers of thankfulness to God continued. And again I wept with tears of joy. Gradually everything began to fade and when everything was almost finished I

After Atheism

Self-styled Cossacks stand guard with flags during an Orthodox service in the temporary structure erected on the site of the death of the last Tsar and his family

felt that quite a lot of time had elapsed, maybe forty minutes or an hour, or perhaps more..."

Yefremov's design is an architectural masterpiece, combining traditional elements into a contemporary style. His plans envisage a complex containing not only the Cathedral church but also a library, icon shop and even a hotel. In a document dated 26th October 1992, the previous archbishop of Yekaterinburg and Verkhotursky, Melkhizedek, had formally confirmed Konstantin Yefremov as the main architect for the project, but from then on the church authorities seem to have done next to nothing about the project.

Yefremov has not yet been paid a single rouble for his work. The 'powers that be' keep putting off the project and claim that they do not have enough money for it.[3] The original plans to build such a church have been thwarted by corruption within the higher echelons of the Orthodox church itself. Donations towards the work have 'disappeared' through theft and now another architect's name is being advertised in the pictures of a proposed new church on display at the site. Meanwhile, a small wooden shelter has been erected next to the simple Orthodox cross marking the site of the Romanovs' execution. The wooden shelter has been used as a chapel for the performance of Orthodox services, including some conducted by Nikon, the former bishop of Yekaterinburg, who had shown little interest in promoting the plans for the construction of a substantial new church on the site of the Romanovs' execution. In February 1994 Nikon had affirmed to Yefremov that the architect's vision regarding the new church was certainly from God, but

Nikon has never bothered to take up Yefremov's invitation for him to inspect for himself the models and plans.

Even within the Orthodox church itself there are those who suspect that Bishop Nikon might have been an appointee of the KGB. Their suspicions are founded on the observations that, prior to his appointment, Nikon had never graduated from a seminary (although he was still a part-time student of theology) and the fact that, after his military service, Nikon had received a 'spiritual education' from Metropolitan Mefodi of Voronezh, whose "overt collaboration with the KGB" was reported in the *Moscow Times* on 3rd July 1998. If Nikon's meteoric rise to eminence had anything at all to do with connections with the KGB, it is not surprising that he should show little interest in the construction of a new church with politically sensitive connotations. In 1999 Nikon was asked to 'retire' (at the age of 39) because of allegations of sodomy, extortion and taking bribes.[4]

Power corrupts

Nikon's negative attitudes towards some other Orthodox priests was vividly shown on 5th May 1998 when, according to an article in the *Nezavisimaya Gazeta*, Nikon ordered their books to be removed from the library of the local Orthodox seminary and burned. The books in question were those by Fathers Alexander Men, John Meyendorff and Alexander Schmemann. It appears as if their writings are being 'banned' elsewhere too: in October 1998 an Orthodox Christian in Moscow told me that she had tried unsuccessfully to obtain copies of Alexander Men's works and had been told by one Orthodox bookshop that "we no longer stock that kind of book."[5] Even though Alexander Men's more 'progressive' ideas are perhaps controversial in some Orthodox circles, Fathers John Meyendorff and Alexander Schmemann were described by Maxim Shevchenko in the *Nezavisimaya Gazeta* as "the greatest Orthodox theologians of the 20th century"; Patriarch Alexei II has repeatedly spoken of his admiration for these theologians, even describing Alexander Schmemann as his "greatest teacher." However, when an Orthodox priest in Yekaterinburg, named Oleg Vokhmyanin, refused to swear on the Bible that he condemned the "heresy" of these three theologians, he received a decree signed by Bishop Nikon banning him from performing his clerical ministry for the rest of his life.[6]

This auto-da-fé in Yekaterinburg was indicative of wider tensions within the Orthodox church between 'traditionalists' and 'reformists', but factors other than theology have also fuelled such tensions. Reformists also suspect corruption among some members of the higher hierarchy who had been robbing the church of valuables for the sake of personal gain. On 17th November 1993 the national newspaper *Izvestiya* had already published an article by M. Smirnov exposing the corruption associated with Nikon's predecessor, Archbishop Melkhizedek.[7] An earlier article had reported the disappearance of Father John Gorbunov, who had stolen a large amount of money donated to the fund for building an Orthodox church on the site where the family of the last tsar was murdered in 1918. It appears that Father John had many contacts with the mafia and criminals in Yekaterinburg. His escape was facilitated by the fact that the archbishop had given him documents allowing him to work anywhere in Russia. Smirnov, a former elder of the church in Yekaterinburg, alleged that some priests — Konstantin Kaunov being named as one of them — pocketed for their own use money which had been given for the restoration of church buildings. He asked whether the theft of money and valuables by John Gorbunov — the vice-president of the above-mentioned fund — could have been achieved without at least the implicit involvement of his superior. In Yekaterinburg many wealthier people had given large amounts of money towards this project and the archbishop knew that Gorbunov had easy access to gold, dollars and valuable icons.

There also remain many unanswered questions about the location of other valuables donated to the project which seem to have 'disappeared'. For instance, a letter signed by Archbishop Melkhizedek had thanked the director of the Maloshevskaya mine for a donation of 20kg of emerald-bearing ore. Three witnesses had seen the archbishop receive the ore but it is not known what he subsequently did with the emeralds. Similarly, it is not known what has happened to the two valuable icons donated to the proposed new church by a wealthy resident of Yekaterinburg named Iliya Karpechko. One of the icons depicted Elizaveta Fyodorovna, the sister of the Tsar's wife, who was afterwards canonised by the Orthodox church. Another unanswered question concerns the $3,000 donated by Germans and Americans for the building of the church where Tsar Nicholas II and his family had been killed. Even though an official receipt had certified that Melkhizedek had accepted this donation, he denied having received the money when questioned about it. On another occasion Melkhizedek is reported to have been given some very old and valuable icons which had been registered as national treasures. The icons had been collected from some small towns and villages in the region, handed over to the church and then taken to the Archbishop's home but were apparently never again seen in public.

Melkhizedek is also reported to have received precious stones, gold, silver and valuable icons from the abbot of the monastery at Verkhoturiye. There are suspicions that these items were taken from old graves but it is a criminal offence to dig up old graves without special permission. If this was the case, was the archbishop aware of the source of the icons?

Smirnov's article in *Izvestiya* claimed that Melkhizedek only visited richer parishes and neglected poorer ones like Sylva or Kachkanar which could not afford expensive payments or gifts for him. On a visit by the archbishop to a local church, the parish was expected to provide a lavish feast for him after the festival. Such feasts would cost at least half a million roubles and included plenty of high quality Smirnoff vodka and other luxuries such as caviar, pineapples, bananas and other rich food. One eyewitness commented that these had to be provided by churches with no paint, heating or proper flooring, and even a shortage of cooking utensils, and where the parish income is largely from poor pensioners.

Did the archbishop also have links with the mafia? Is it true that in his chapel at home the archbishop baptised people known to belong to the mafia and who were known to have gunned down people in cold blood? Did they pay the archbishop well for his services? Was one of the mafia members also made a deacon — with an income from the church? Those involved with the church noticed how some young men who sung well in the choir were given positions of authority in the church — but were often drunk all day. Coffins of those believed to be mafia members, who had been killed, were seen outside the church awaiting burial: one local Christian observed that even the mafia have a desire to preserve their souls if they believe it can be done simply by paying enough money!

These suspicions about apparent avarice and possible corruption in the former archbishop of Yekaterinburg might help to explain why parts of the church hierarchy had been so opposed to Father Gleb Yakunin — a priest who also became a member of the Russian duma. In an open letter to the patriarchate, Father Yakunin had named Archbishop Melkhizedek and some other archbishops (Serapim from Tula, Mefodi from Voronezh and also Archbishop Gideon) as being responsible for the loss and destruction of church property. The patriarch's response was addressed not to Yakunin but to the state duma, requesting the deputies to relieve Yakunin of any responsibilities for religious issues. This was prior to the patriarchate's decision to find a rather contrived excuse to defrock Yakunin because he was involved in politics.[8]

Informants for the secret police

The adage that 'power corrupts' is unfortunately true not only of some people with positions of power or influence in government or industry but also within some religious organisations. Of course there are also those who remain true to their moral standards, but sometimes the temptation to compromise one's values can creep up slowly and insidiously. Such pressures occur in all religious groups, not only in the Orthodox church, although I suspect that the temptations might be stronger in those organisations which have established themselves in positions of power and influence, especially among those who are in senior positions within their hierarchies.[9] Usually these are men who had been appointed to their positions during the Communist period, when the senior hierarchy was considerably influenced by the KGB. Promotion to senior posts was almost invariably made with the consent of the KGB, who made sure that their own 'puppets' were given such posts at an appropriate time (even if they had to wait in junior positions for some years before their promotions).

Those who had started off as KGB informants or 'plants' at first might not have felt that they were in any way betraying their faith. For instance, when Tanya Osintseva had been working as Archbishop Mekhizedek's secretary she had been summoned to the KGB on three occasions. At the first meeting she was asked if she was a patriot, to which she replied: "Of course I am a patriot." It was on the basis of her loyalty to her country that she was then asked to let the authorities know about the activities of the Orthodox church. Tanya believed that her loyalty to God had to take precedence over her loyalty to the state so she refused to become an informer, but it does appear as if many of those who had been appointed to higher positions in the official hierarchies of different religious organisations are those who had been willing to express their patriotism in the form of regular reports to the KGB about others in their organisations. This is reported to be true not only of the Orthodox and Muslim institutionalised hierarchies but also of some members of the officially recognised Council of Evangelical Christians and Baptists during the Communist period; one such man whose colleagues thought he was probably a KGB informant was even described by them as "an honest man — because he refuses to take communion with us"!

Nowadays it is unclear to what extent there is continuing surveillance of religious organisations by the FSB, the successor to the KGB.[10] Certainly it is far less overt. What might be the tip of the iceberg was revealed to me in 1992, when I took a photograph of a church service in Almaty, Kazakstan. Afterwards a man came up to me, showed his KGB card and asked why I had taken the photo. I replied that it was simply out of interest and asked why he wanted to know. Recognising my accent as not being that of a native Russian, he asked where I was from and was apparently satisfied when he learned that I was English.

Humanitarian aid

Sometimes Western finance is accused of giving Protestants an 'unfair' advantage in relation to the Orthodox church, but the above-mentioned allegations of corruption in some of the higher echelons of the Orthodox church give a different perspective to this issue. It might not be a question of the source of funds but of the right use of available resources. From what I know of many situations throughout Russia, the majority of both Protestant and Orthodox churches are supported primarily through the donations of their parishioners, with only a relatively small proportion of their income coming from elsewhere. When foreign help is available it often takes the form of sponsorship of specific short-term projects. One form of co-operation is that of the provision of humanitarian aid for those in need.

Some Orthodox Christians have also become actively involved in social concern and humanitarian aid of one form or another. For instance, Tanya Osintseva took me to visit an orphanage and also a hostel for abandoned children in Yekaterinburg. I had brought with me children's clothes, toys,

The Orthodox kaleidoscope

medicines and other forms of aid donated by people in the West. The director of the orphanage told me the stories of some of the children, such as Natasha, a girl of about eight years of age with beautiful blonde curly hair. She was found abandoned by her mother, left behind in the room where they used to live. Until her mother disappeared, Natasha had been the one who supported her mother and younger brother by begging. She knew what to say to people when they answered the doors to her, using very subtle psychological techniques to persuade people to give to her.

Another Natasha was six years old. She was brought up by her great-grandmother because Natasha's own mother and grandmother had both become alcoholics and could not look after the little girl. Finally the great-grandmother came to the conclusion that she herself could no longer manage to bring up Natasha. She obtained a medical certificate to verify that at her age she was physically unable to look after a young child like Natasha. Then she took the certificate to a state orphanage and asked them to look after the child instead.

"You have to look after yourself now," 13-year-old Lyuba was told by her mother in February 1994. The mother then left with her five younger children for an unknown destination, eluding all efforts by the authorities to trace her. Lyuba was left in the flat all alone. At first she found food but then she lay down and just wanted to die. The neighbours discovered the situation and came to fetch Lyuba. She ended up at this orphanage, where the staff discovered that the girl had never been to school and could neither read nor write so she was given a special educational programme which has enabled her to become literate. Now Lyuba studies and goes to school, often receiving special additional tuition from the teacher. She also works in the orphanage kitchen: her greatest desire in life is to be a cook.

Nastya was a little girl whose age was estimated by the doctor to be about five years. In November 1993 she was found in the street but her parents could not be located despite information given out about her over the television and radio. In the absence of other documentation, the orphanage had to get her specially registered with the authorities. Nastya sees all right but has a squint which puts off those wanting to adopt a child. She still longs to be reunited with her mother.

Anton was a seven-year-old boy who was admitted to the orphanage on the morning we visited it. He'd been there only half an hour and remained generally silent, not speaking about his mother or where he had come from. He too had been found in the street and nobody knew anything about his parents.

The father of Sasha, another little boy at the orphanage, was known to be in prison. Sasha's mother took him to the orphanage and abandoned him. When they tried to trace her, it was found that she had left her former home and had vanished without trace.

None of the children in this orphanage were true 'orphans' with no living parents, but all had been abandoned by their parents. For instance, a boy named

After Atheism

The hostel for 'street children' in Yekaterinburg. The boys are sitting on benches around the walls listening to Tanya Osintseva telling them about Jesus

Alyosha has a mother who sleeps under a bench in the botanical gardens. The alcoholic mother of an eight-year-old boy named Gena and his younger brother Ruslan sometimes forgot to feed her children while she was drunk. Their father was then given custody of the children, but later he threw Ruslan out of the window of their flat on the fourth storey of an apartment building. Now the father is in prison and the boys have been taken into state care.

Over two thousand children are abandoned each year in the central Urals region of Russia — far more than the mere 23 children who could be accommodated by this orphanage on the outskirts of Yekaterinburg. By the time of my first visit in 1995, only nine children from the home had been adopted; a further six were with foster parents. Although the children remembered their real parents and wanted to be with them again, often they were embarrassed to talk about their parents openly.

Elsewhere in the city there is a special hostel for boy 'street children' who have been found sleeping out at the railway station or in parks or other public places. They stay at the hostel for an average of about three months or so before either being reunited with their parents (if the authorities can trace them) or else being moved to another kind of institution. In the hostel there is little or no real schooling. Most of the boys' time is spent watching television or playing games. A few of them have been found work to do around the hostel, such as cleaning floors or ironing sheets. Those we spoke with in the laundry room said they preferred doing the ironing because it at least gave them something constructive to do.

The Orthodox kaleidoscope

However, in the whole city (which has a population of 1.5 million) there is no such hostel for girls. In the early 1990s the Yekaterinburg authorities began to construct a girls block at their hostel, but building work stopped when there was no longer enough money to pay for the project. The nearest girls hostel is in Nizhnye Tagil, a few hours journey to the north, but it does not have the capacity to accommodate all the girls from Yekaterinburg. In the absence of any other facilities, some healthy children have been accommodated in local hospitals if there was space. However, sometimes girls found sleeping out in the streets have been left there because the police were unable to find a place for them to be accommodated. The fear is that such girls easily end up as prostitutes: a Pentecostal pastor cited a statistic that already there were at least thirty brothels in the city.

In Russia as a whole there is an increase in homelessness among children on account of the economic chaos. Some impoverished parents feel unable any longer to look after their own children, so they suddenly move to another city and abandon their children. Often the children have had to steal in order to find food while they were sleeping rough. At the orphanage I was told that they had received help from Protestant groups like the Salvation Army and the Pentecostals but Tanya was the only member of the Orthodox church who had been motivated enough to try to help.

On two occasions Tanya Osintseva also arranged for me accompany her on visits to the Yekaterinburg central prison. It was designed for 2,500 prisoners but in fact contains about 6,500. However, the government now only allocates money for it on the basis of its official capacity of 2,500, and that budget was itself being slashed. As a result of this policy, many of the inmates were very malnourished and thin — in fact, half starving.

Many of the prisoners were not actually convicted criminals but had been imprisoned while awaiting their court hearings — which in some cases could be three years after their arrest. I was told that during this time they were not even allowed to write letters to relatives or friends. Quite a number of the prisoners were teenagers who in some instances had been imprisoned for what appeared to be relatively minor offences; examples included thefts of a bicycle, a box of vegetables and even a case of a boy being detained in prison after he stole two jars of jam. However, in the prison they mixed with hardened criminals, about a third of whom had been arrested on murder charges.

At the time of my visit in 1995, there were 320 cases of tuberculosis in the prison and also an outbreak of diphtheria. However, they were very short of medicines with which to fight these diseases. They also needed analgesics and medicines for gut infections, skin diseases, heart problems, anti-lice medicines and sanitary materials such as soap, washing powder and disinfectants. On a visit to a cell for more hardened adult criminals, a warder told me to come away from the bunk beds on which I was leaning so I would not be infected by lice! I had brought with me a selection of medicines, including erythromycin to combat diphtheria in carriers of the disease, which we donated to the

After Atheism

prison doctor. (I asked Tanya to check up that they were actually put to good use.)

Previously Tanya Osintseva had regularly been visiting the prison as an assistant to Father Vladimir Domrachev, a young Orthodox priest who had a genuine concern for needy people and for those who feel that nobody cares for them.[11] A sense of rejection can sometimes turn into bitterness and anger against the society in which they have not felt accepted or wanted. However, those who do try to show love and genuine concern are often motivated by the love of God. The fact that many of the prisoners recognised and

Prisoners at an overcrowded prison in Yekaterinburg

responded to the concern of these Orthodox Christians is reflected in the statistic that over nine hundred prisoners asked to be baptised during the one and a half year period when Father Vladimir was able to visit the prison. However, he had to cease his visits after he was reassigned to a remote parish in the village of Kardapolye in the Kurgan region.

From hippy to Orthodox priest

Another Orthodox priest who has had an effective prison visitation ministry is Boris Razveyev. He belongs to a new generation of younger Orthodox priests who had sometimes been themselves victims of the KGB and who now seem to have a greater pastoral concern for the welfare of their parishioners. When I went to visit Boris in Ufa, a friend commented: "If you get accosted by any of the street gangs, tell them that you're on your way to visit Father Boris, because they know him from his visiting the prisons." I'm glad I did not need to try putting his advice into practice, but it showed me the high respect for Boris Razveyev among some of the criminal gangs in this part of the Russian Federation.

In the 1970s, however, Boris was so looked down on by some of the city's population that they threw tomatoes at him! That was when he was a hippy and such people were not normally seen in this provincial city. Boris showed me photographs of his appearance at that time — quite a contrast to his present attire as a Russian Orthodox priest!

Boris has baptised over eighty people in prison. He has a special letter from the local church hierarchy granting him the right to visit those prisoners who have been given the death sentence. Boris not only speaks to them about Jesus but he also does what he can to bring them practical help such as bringing in for them items such as packets of tea. He well understands the feelings of prison inmates because he himself had been imprisoned for thirty-two months from 1984 to 1986. In fact, Boris had even been born in a prison camp and had spent the first six years of his life (from 1949 to 1955) in a labour camp in the cold Magadan province of north-eastern Siberia, where his parents were prisoners. While in detention, Boris' father had met the son of the dissident Peter Yakir, who in 1974 visited the Razveyev home in Ufa. This attracted the attention of the KGB, who wanted to know why this famous dissident had made a telephone call to Ufa. Shortly afterwards, Boris attended the trial in Moscow of Yakir and another dissident named Krasin. A police officer asked him why he was there, examined Boris' internal passport and wrote down his name and address. From then on the KGB were interested in his activities.

An experience which changed his life occurred in October 1972. Boris was in an armchair listening to music by Bach when "suddenly I felt as if I was levitating from my chair. I had a vision of a large church or cathedral but nobody was in it. There was a lot of dust and a window was broken but a ray of light came through it and fell on the altar. Immediately people began to sing — I saw a choir and a church full of kneeling people. Their sins came out and they became clean. Then a woman appeared dressed in white with long sleeves. She took me by the hand and I saw myself as a small boy with fair hair. She was smiling and joyful. Then, with me in her hands, she glided through the wall out of the church and to a lawn in front where she stopped and I lay on the ground. She began to stroke me and I was sleeping. Then it was all finished." Commenting on his experience, Boris remarked: "For me it was a sign. I was an atheist and had not had any previous experiences of this kind — no transcendental meditation or anything like that. It was my first experience of a 'metaphysical world' and so then I began to get interested in philosophy, religion and so on but it was two years later that I became a Christian."

On 26th July 1975 Boris was baptised in a kitchen by Father Dmitry Dudko.[12] After his baptism, Boris felt that he was a "new person" and that "a whole new world" had opened up to him. In Moscow he also attended the Christian seminars organised by Alexander Ogorodnikov, a well-known Orthodox Christian. On returning to Ufa Boris began to go to church but was soon cross-examined by the KGB. Boris later co-authored with Ogorodnikov a special letter which appealed for help for Christians in Russia and provided a considerable amount of "real" information about the situation for believers. Boris paid for this act when the KGB then prevented him from continuing his studies in law at university. His student identity card was not confiscated but when he tried to use it to obtain a discounted ticket it gave the KGB the pretext they wanted to have Boris arrested. For such a relatively minor offence he was imprisoned for thirty-two months, commencing on 31st January 1984, but it was only because of Gorbachev's more liberal policies that Boris was released towards the end of 1986. He recognises that the real reason for the imprisonment was his Christian faith — as confirmed by the officers taking him to prison who told

Boris Razveyev

The Orthodox kaleidoscope

him he would stay there for many years because the government preferred him to be out of the way for a long time.

Almost every week while he was in prison there were sessions of political indoctrination. However, Boris announced that the lessons were not for him because he was an anti-Soviet prisoner and he would prefer it if the lessons started with prayer! He was then put for three days in a special internment penalty room in the middle of the camp — a very small cell with no bed.

Some might regard Boris as a hero but some Orthodox believers regarded Boris as a traitor because they remember an incident during the Communist era when, in a publicly televised interview, Boris admitted his involvement with the underground Orthodox church. Apparently his confession was extracted by a KGB threat of imprisonment, but some believe that it resulted in others being arrested instead.

More recently, Boris had a happier meeting with the KGB, who told him that they were closing their file on him. They gave him a copy of their report on him, which, among other things, details some of the young people whom Boris had influenced towards a Christian faith. (One of them was later put into a psychiatric hospital, where the doctor rather boldly wrote on his notes that he was confined "by order of the KGB.")

Boris had wanted to be a priest but had been prevented from doing so by the Communist authorities. For eight years he had tried each year to study at a seminary, but his applications had been constantly refused by the KGB, who did not want younger people or dissidents in the ministry. In 1990 he was wondering about trying to emigrate to America so he could become a priest there, but then the former bishop of Ufa himself emigrated. His successor then helped Boris to become a priest. Now Boris himself is helping to train others who want to go into the ministry.[13]

Top: Boris Razveyev conducting the liturgy at his church (as seen from the priests' side of the iconostasis). Bottom: Preparing the communion bread

A marked contrast with the predominantly elderly congregations of most Orthodox churches is the fact that younger people, from all walks of life, constitute about three-quarters of the congregation at the Church of the Birth of Our Lady, where until recently Boris had been the ministering. Many complete families attend, and about a third of the congregation are schoolchildren, students or recent conscripts into the army. Between 1991 and 1993 the average Sunday congregation had risen from about 35 people to 450, increasing to between 700 and 1,000 people for special festivals. To some extent this dramatic rise in church attendance was due to a series of events in the autumn of 1992 which the Orthodox church has now officially recognised as miraculous. Many remarkable healings were reported when people were anointed with a mysterious oil which for three months was exuded by some of the icons in the church, as will be discussed in more detail in chapter 11.

This church had been used as a cinema during the Soviet period and was in a dilapidated condition when it was handed back to the Orthodox clergy. While Boris was there the parishioners worked hard at restoring the building, involving clearing away debris by hand and laboriously rebuilding the tower without the help of cranes or other sophisticated machinery. They had almost finished the work when the president of Bashkortostan, Murtaza Rakhimov, took an interest in the project and decided that it would be a suitable venue for official State ceremonial functions. (Despite being nominally Muslim, Rakhimov still attended Orthodox festivals in his official capacity as head of Bashkortostan.) Rakhimov offered to allocate State funds to finish off the building project but he did not want Boris to remain as the priest in charge because it did not seem 'fitting' for someone with a 'criminal record' to officiate at events attended by the President and other dignitaries. Therefore Rakhimov asked the Patriarchate to reassign Boris to a position elsewhere. In view of these circumstances Boris felt that he did not want to remain in the Ufa diocese so the Patriarch offered him a position as priest in charge of the church at the Vagankovskiy cemetery in Moscow.

In July 1998, three months after he had moved to Moscow, I visited Boris at his new apartment and ended up helping him and a few younger priests to assemble furniture. He said the move to Moscow had been difficult after all his time in Ufa, when he had formed close ties with his parishioners. From his religious perspective, he now had to ask why God might want him to move with his family to Moscow and to start afresh.

An Orthodox house group

Orthodox Christians elsewhere in Moscow had also made a new start in the way in which they were expressing their Orthodox faith. I attended one of their meetings in a Moscow apartment where a dozen people — on this occasion, six men and six women — had gathered together. At the beginning everybody prayed for the young man whose turn it was to lead the Bible study that week.

The Orthodox kaleidoscope

He knelt on the floor while the others prayed, some of them putting their hands on his shoulders. Then he led them in a Bible study; that week's passage was from the Acts of the Apostles as they systematically worked their way through the book. Each person present was encouraged to contribute to the ensuing discussion of the passage.

The Bible study was followed by a time of prayer during which Russian Orthodox practices became a little more apparent. Everyone stood in a semi-circle facing the icons and cross which were above the hearth on one wall of the room. Anyone who wished to pray did so, without any set order or liturgical pattern, either in Russian or another language. During this time of 'open prayer' several people stood with their hands raised upwards, but sometimes during prayer they might also cross themselves in the Orthodox fashion.

An invitation was then given for anybody who wanted special prayer to ask for it. A woman named Lena said she had been struggling with depression. She knelt on the floor, facing the cross and the icons, while all the others gathered round her, several of them laying their hands on her head and shoulders. A number of people prayed for God to minister to her. Then another lady, named Karina, asked for prayer for healing of a neck pain. She too knelt while the others prayed for her. During the time of ministry one of them appeared to have received a revelation from God which he whispered in Karina's ear. She smiled and said: "Thank you."

The evening concluded with supper together between about 10pm and 10.30pm. Tea was served, and on the table were put out bread, cold meat, honey, nuts and a bar of chocolate broken up into separate squares so that everybody could help themselves to what they wished. I learned that several of those present had become Christians through their contact with this Orthodox community. For those who find the archaic language of the liturgy hard to understand, and fail to see the relevance of the church for modern life, these kinds of Orthodox house fellowships are important in bridging the gap between the traditions of Orthodoxy and practical issues of everyday life.

Seven different Orthodox house groups connected with this same Orthodox congregation meet every Monday evening in various parts of Moscow. To prevent each group becoming too much of a self-contained clique, there is an occasional shuffling around of the members so that they all know people in the other house groups. During their times of open prayer they often pray for those in other groups. Before the time of prayer, the names of such people are distributed on slips of paper so that each participant has one other person for whom they will be praying during the coming week.

Members of these house groups are active in the Orthodox church which they attend on Sundays, but the house groups and some humanitarian aid activities are organised by a fellowship within the Orthodox church which goes by the name 'Hosanna'. Two of the active participants in the Hosanna fellowship told me that for many years they had attended a prayer group organised by the Orthodox priest to whose parish they belonged. These prayer groups provided a basis for further lay participation in corporate spiritual activities beyond their regular attendance at the liturgy. Prayer continues to be a core element in the Monday evening house groups, which have now developed into a fellowship which the participants regard as a kind of community in itself.

Over fifty people attend the regular Monday evening meetings and are regarded as members of the Hosanna community, but often they meet together at other times during the week for different kinds of activities. One of these is a ministry to the poor. The community helps to support seventy poor families in Moscow, some of which have large numbers of children and are unable to 'make ends meet' in the current economic climate. As the community does not have enough resources in itself to help all these families, they are grateful for the help of Orthodox friends in France who have sent money for this work.

A time of prayer for members of the Orthodox Hosanna community

Since 1992 the Hosanna community has also organised a special "school of faith" for newcomers to their parish. These classes teach the elements of the Christian life and of fellowship with God. Their hope is that this might become a model for other parishes too. Special parish 'retreats', meeting in a large hall, have also been organised with the aim of challenging parishioners about the call of Christ upon their lives as they consider the many needs around them in the parish, the city and the country. In 1993 they organised two retreats — one before Easter devoted to the passion of Christ and the other before Pentecost about the Holy Spirit.

One problem they are seeking to overcome is people's lack of familiarity with the scriptures. Therefore they have already published several books, including a modern songbook for young people and 25,000 copies of a booklet entitled *Meet Jesus*, intended for children and mentally disabled people, because it contains many pictures and little text. This booklet has proved to be so popular that it is now virtually impossible to obtain a copy! They also published cards reproducing an icon which depicts the apostle Andrew (representing the Orthodox church) embracing his brother Peter (representing the Western, and particularly the Catholic, church). In January 1964 this icon had been presented by Patriarch Athenagoras of Constantinople to Pope Paul VI as a sign of reconciliation between these two branches of Christianity.[14]

Despite this desire for reconciliation among many ordinary Orthodox Christians, a reluctance to develop relationships further with the Roman Catholic church was expressed in 1998 by Patriarch Alexiy II, who said it was still too early for him to meet Pope John Paul II "for what would be the first top-level talks between the two Christian churches in a millennium." In written answers to questions, Patriarch Alexiy II is reported to have cited two

main problems in relations between the Orthodox church and the Vatican, namely the treatment of Orthodox believers in the western Ukraine and Catholic proselytizing inside Russia. The patriarch's five-page document stated: "The second problem is the persistent missionary activity of various Catholic missionaries among people baptized as Orthodox with Orthodox roots. They have allocated substantial resources to such activities in Russia and other countries of the Commonwealth of Independent States, involving hundreds if not thousands of priests, monks and lay people. Until there is serious, clear progress in solving these two problems, the majority of Orthodox cannot positively accept my meeting the Pope."[15]

Rhetoric of this kind seems to overstate the degree of Roman Catholic involvement in Russia, most of which is merely a re-establishment of ministry among dispersed Roman Catholics whose churches had been confiscated by the Communists or else were not allowed to register as parishes.

The Orthodox debt to Protestantism

Perhaps the patriarch does not realise how much his own church has actually benefitted from the activities of other denominations. In several cities — including Yekaterinburg, Ufa, Izhevsk and Moscow — I have come across people who became Christians through contact with Protestants but later began to attend Orthodox churches. If they are at all representative of the 'active' members of Orthodox churches, it seems as if the Orthodox church owes a great debt to Protestant preachers.

A Bashkort friend of mine, for instance, first became a Christian at a charismatic Protestant church but he later chose to attend an Orthodox church because some of his friends went there. I also know of Orthodox Christians who first came to faith through reading literature written by Protestants. One of them is an Udmurt pensioner who commented that her life had been changed through the books she had read, and she wished such literature had been available to her earlier in her life. In another instance, a Russian young woman studying at the Urals State University had needed to look up the biblical references in the chapters of a book which she was translating from English as part of her coursework. Therefore she went to the university library late in the evening to read the Bible when nobody else was around to see what she was reading. This was the first step in her coming to faith but she now regards herself as Orthodox despite having read Protestant literature and having attended at least one Protestant meeting.

At a charismatic church in Bashkortostan I met a Tatar woman who had felt unable to talk with the pastor about a moral dilemma she was facing. She was pregnant by her boyfriend who was making her have an abortion because she already had one child and her boyfriend did not want a second child. She felt trapped by her circumstances, knowing that her pastor would not approve of either her intended abortion or of her still living with her boyfriend without being married to him.[16] On my next visit to Ufa, seven months later, she was living with her brother instead of the boyfriend but was regularly attending an Orthodox church instead of the Protestant one. At her brother's home, she showed me a book she had been studying which explained the symbolism and meaning of the Orthodox liturgy — which many find difficult to understand. Only the more committed Orthodox believers are willing to work through such explanatory books.

Moreover, many people are unable to understand the Orthodox liturgy because it is in Old Slavonic, whereas the Protestant church worship is in modern Russian. The relative simplicity of Protestant services can be attractive to many people, although they can also be put off by some Protestant religious jargon. Both Orthodox and Protestant services often last for two hours or more, but in an Orthodox church it is easier for visitors to arrive and leave whenever they wish while the liturgy is being performed. On the other hand, most Protestant churches provide seats (unless, as sometimes happens, the meeting is so crowded that there is standing room only) whereas Orthodox services require the participants to stand all the time except if kneeling in prayer at the altar rail.

These kinds of superficial organisational features only partly explain personal preferences for either an Orthodox or a Protestant form of church service. Probably far more important is the social context in which a person is introduced to the church. Very often those who are invited to a Protestant church service are afraid that they would not know what to do and are embarrassed if they do not know what to say during a time of prayer. It is a

The Orthodox kaleidoscope

For those who live in impersonal apartment blocks (such as these in St Petersburg), a church can provide a community in which social relationships are more meaningful

reassurance to them if a friend accompanies them and explains what is happening. Even though Orthodox services use archaic language and forms of ritual, there is nevertheless a feeling that one can walk in or out whenever one likes: in this sense, it is less 'threatening' because it involves less personal commitment.

Nevertheless, in both Orthodox and Protestant churches those who become more committed tend to be those who have found within the church a network of friendships with people they feel they can trust. The discovery that there are people who share a common commitment to certain ethical standards and have a love for one another is an attractive aspect of the Christian church for those who sense a lack of trust in many aspects of their ordinary social relationships.

There are also features of many Protestant churches which have been influenced by Russian Orthodoxy as well as by Western practices. Some older Baptist churches have a relatively rigid, male-dominated hierarchy which seems to parallel the Orthodox hierarchy but is regarded by the Baptists as a biblical model. In most Orthodox, Baptist and Pentecostal churches women have to wear head coverings and are forbidden from using cosmetics or wearing tights. To some extent, such attitudes are also consistent with Protestant German pietistic influences prior to the Communist period, but in Russia they are reinforced by the fact that similar practices are prevalent in the Orthodox church too. One of the reasons why younger people in Ufa were attracted to the ministry of Boris Razveyev was the fact that he allowed women to wear make-up during an Orthodox service!

Nowadays many people who are attracted to Christianity in terms of its ethical standards or other teachings are put off by the excessive legalism in both Protestant and Orthodox churches. In a culture in which drinking alcohol and smoking are very widespread, new converts are often put off by the rigid teetotalism and prohibitions on smoking which are common in many Baptist and Pentecostal churches. Some observers think that the Protestant churches would probably attract ten times as many people if they did not have some of these regulations.

Attitudes are changing, especially among some of the newer charismatic churches. Their greater freedom in worship and use of contemporary styles of music is attractive to many younger people, but can be offputting to older people. Some Orthodox Christians also like the more modern types of songs which are used in certain Protestant churches. Superficial differences between churches are similar to the kinds of variations which occur between cultures: one culture is not 'better' than another, merely different. There is a strength which comes from diversity whereby the various denominations — including the Orthodox church — can appeal to different generations and various kinds of temperaments.

As the 21st century dawns, there is an increasing recognition not only among Protestants and Catholics, but also within some sections of the Orthodox church, that the church has to 'move with the times'. Increasingly a distinction is being drawn between 'form' and 'substance', whereby the basic truths can remain unchanged but the forms in which they are conveyed need to be sensitive to the culture of the audience. For instance, the music of an Orthodox rock group named Galactic Federation has helped many younger people to find faith within the Orthodox tradition.[17] Different forms of change are appropriate for various contexts, so some Orthodox parishes which use modern language for the liturgy (instead of Old Slavonic) still use traditional vestments and styles of music, which probably give visitors a greater sense of security and familiarity. Generational differences are almost tantamount to differences between different sub-cultures, so there may be a need to express the truths in ways which are appropriate to that sub-culture. In virtually all denominations — including Russian Orthodoxy — there are 'traditionalists' who want to preserve the forms of a bygone age, and 'modernists' who want to make them more relevant to the contemporary generation.

To some extent this spectrum within Orthodoxy also reflects the degree of openness to meaningful interaction with other denominations, beyond the formal level of 'official recognition' or membership of the World Council of Churches. It is this level of religious expression through formal organisations which is the most obvious manifestation of religious behaviour to an outside observer, as compared with personal spiritual experiences such as dreams and

The Orthodox kaleidoscope

visions or the forms of religious behaviour which are prompted by crises such as the death of a loved one. Moreover, the diverse forms of Orthodoxy outlined in this chapter are not only to some extent religious expressions of what it means to 'be a Russian' but are also expressions of personal taste and of 'sub-cultures' among the Russian people. Other religious sub-cultures exist too, but outside the Orthodox church — which poses a challenge to those who view Orthodoxy as the only legitimate expression of what it means to be religious and to be Russian too. Some of these other forms of organised religion will be examined in the following chapters.

8
Free market religion

"**R**ussia is holy — Preserve the Orthodox faith!" "An error has come: here live the Orthodox!" "God is not a Father to the one to whom the Church is not a Mother!"
With such eye-catching slogans on their placards a group of protesters demonstrated in front of a Russian sports hall while a Protestant mission was being conducted inside. Those trying to pass the demonstrators to get into the building were given anti-Protestant leaflets. Several of the protesters were self-styled 'Cossacks' whose grandparents had been in Cossack regiments disbanded by the Communists — even though many Russians in the city think it is ridiculous for the descendants of such people to wear a Cossack uniform and to promote ultra-Orthodox and nationalist values. Nevertheless, these 'Cossacks' were legitimately exercising their right to a free expression of their beliefs.[1]

Had the demonstrators cared to come inside, however, they might have been surprised to hear the Finnish evangelist Kalevi Lehtinen encouraging people to attend not only Protestant but also Orthodox churches. Towards the end of each evening's meeting, he would say: "We are not trying to make Orthodox people into Protestants, or Protestants into Orthodox: we want people to follow Jesus. You know where the Orthodox churches are, and you can go there if you like. If you want to go to a different kind of church, we have a leaflet here with the names and addresses of participating churches."

After one of the meetings, a man came up to me and asked where he could find a church of the 'Old Believers'; I gave him the address and telephone number of an Orthodox friend of mine who could help him. We do not know how many of those who went to the Protestant meetings then started to attend Orthodox churches. Many of them would have felt more at home with an Orthodox style of service. However, some of the Protestant pastors were disappointed that their churches did not see as many converts as they had hoped. Some people who did respond refused to give their names and addresses because they had been so used to fearing the KGB that they still did not want to divulge such details to strangers.

An Orthodox journalist named Oleg who participated in the mission commented that this form of mission is difficult for Orthodox Christians to understand. It took Oleg a while to get used to it, but now he sees how important it is for the different denominations to work together. Other Orthodox Christians were also actively helping this Christian mission

Free market religion

Placards used by Orthodox demonstrators in Yekaterinburg

organised by eight different Protestant churches in this city of Yekaterinburg.

Pavel Bak, the pastor of a flourishing Pentecostal church in Yekaterinburg, afterwards estimated that about 250 people filled in response cards as a result of this mission. During 1994 and until the end of March 1995 a series of such missions were conducted in over a dozen different cities in the Urals region, including major centres such as Yekaterinburg, Chelyabinsk, Izhevsk, Perm, Orenburg and Ufa, as well as in several smaller towns. On the whole, the organisers felt that their missions had a greater impact in the smaller settlements where people were less used to foreign evangelists.

The organisers of this 'Mission Ural' tried to co-operate with the Orthodox church wherever possible. For instance, in the town of Shadrinsk they donated their left-over books of scripture passages to Father Mikhail, the local Orthodox priest. In some places they did meet with opposition, the most extreme instance being in Nizhnye Tagil, a town to the north of Yekaterinburg, where Orthodox extremists had made in the centre of the town a public bonfire of the Protestant literature handed out during the mission. In Yekaterinburg Kalevi Lehtinen made several attempts to contact and meet with Archbishop Nikon but each time the Archbishop found reasons to avoid Lehtinen's request for a meeting together.

However, other Orthodox Christians in Yekaterinburg regarded the hierarchy's stance as bigoted and narrow-minded. Their more open stance towards other denominations was demonstrated three months later through an inter-denominational charitable effort at Christmas to help orphans and poor people. This project, which involved Orthodox, Catholic and Protestant churches, was widely publicised in the local press and the positive tone of the articles showed how the scheme obviously met with public approval. A dozen local and regional newspapers carried headlines such as: "It is more blessed to give than to receive"; "Hurry to do good"; "When charity begins"; "Mercy is greater than religious arguments"; "Joy and hope at Christmas"; "Pastors have unity under the Christmas tree."[2]

Two things impressed the journalists. Firstly, the churches were visibly doing something practical on behalf of the city's orphanages, old people's homes, hospitals, disabled children and the children of refugees. Secondly, those belonging to several different religious organisations were actually co-operating together in a city where, just three months previously, ultra-Orthodox nationalists had been publicly demonstrating against the Protestants.

An Orthodox priest named Vladimir Domrachev was one of the initiators of an inter-faith charitable project over the Christmas period — starting on 23rd December and continuing until 5th January, the Orthodox Christmas. With the help of Tatyana Tagieva, who works for the local government's Department of Religious Affairs, a series of children's concerts, parties and nativity plays were organised throughout the city, for the benefit of about 3,000 people. Christmas presents such as apples and chocolate were given out too. More than 120 volunteers, half of them children, took part, representing eight different religious organisations.

The Protestant participants included the Pentecostal 'Living Word' church, which for several years had already been active in humanitarian relief and visiting old people's homes, prisons and orphanages. Alongside them were both the Baptists (who are often considered to be very conservative and traditional) and the charismatic 'New Life' church. All these Protestants had reservations about also inviting sects such as the Mormons and even the Krishnaites to participate too, but in the end agreed to do so as a way of counteracting the public's impression of religious people as intolerant and bigoted — the very impression which had partly been fostered by the way in which some of the higher clergy of the local Orthodox church had been attacking the Protestants as heretics. The Orthodox upper hierarchy abstained from participation in these charitable events at Christmas so the initiative was taken by Vladimir Domrachev and lay Orthodox believers. They wanted to try to heal the wounds between denominations through a project to which all felt they could contribute in some practical manner. Those divided by theology could nevertheless come together in charity at Christmas.

One of those helping Vladimir Domrachev was Tanya Osintseva, who had already been helping Vladimir Domrachev with his prison visitation ministry and orphanage work. In this she had fostered links with Protestants, especially with the Pentecostals, who respect Tanya as a fellow Christian within the Orthodox church.

Not only some members of the general public but also some Christians within the Orthodox church itself have been suspicious of older Orthodox priests who are assumed to have collaborated with the Communists. However, there is also a widespread respect for the Christian faith of certain Orthodox priests like the late Alexander Men or Dmitry Dudko whose works are read by Protestants as well as by Orthodox believers. In fact, one of Dudko's books has been published and distributed by the German evangelical mission Licht im Osten.

Even though many Protestants are prepared to work together with the Orthodox church if at all possible, they are at times rebuffed by the attitudes of some very conservative priests who will have nothing to do with Protestants. Other Orthodox priests and lay people, however, are willing to work closely with those of other denominations, although sometimes they have had to be discreet about their contacts: for instance, when a Catholic priest was visiting Yekaterinburg, Tanya Osintseva was forbidden by the Orthodox hierarchy from meeting with him in her official role as the archbishop's secretary, but this did not stop her from going to hear him speak in her private capacity, as simply a member of the general public. Many Orthodox Christians also appreciate literature written or published by Protestants, such as *The Chronicles of Narnia* by C. S. Lewis — which I have seen on sale in an Orthodox church in Moscow. Not only lay Orthodox Christians but also some of the priesthood have been using literature consisting principally of readings from the Bible, such as the booklets and leaflets produced by the Scripture Gift Mission, because these are denominationally neutral and are designed to allow the Bible to speak for itself.

To a large extent, the Orthodox faith is regarded as the religion of the Russians, although it of course also embraces a number of other non-Russian peoples. However, the ultra-conservative Orthodox demonstrators seemed to believe that Russians should only be Orthodox and that all other religions are heresies. For them, ethnicity and religion were regarded as virtually coterminous. Since perestroika, divisions within the Orthodox church have become much more obvious. Within Orthodoxy, there is a wide range of opinion and practice, but as a rough generalisation — to which there are of course exceptions — it might be said that those priests who were part of the official establishment during the Communist period have tended to remain conservative whereas many younger priests (like Vladimir Domrachev or Boris Razveyev) are more open to change.

Islamic debates

After the freedoms of perestroika and glasnost in the late 1980s facilitated a widespread interest in religion, initially there seemed a potentially unlimited religious 'market' in which there was room for almost anybody. After the boom period, however, growth began to slow down and some converts began to 'shop around' and try other denominations, though often remaining within the general framework of 'Christianity' or 'Islam'. In this context occasional tensions arose between religious groups within Christianity and also within Islam in terms of a jostling for their 'share of the market'.

During the Soviet period, Muslims of the Russian Federation were under the jurisdiction of the Spiritual Directorate based in Ufa but conflicts among the Muslims led to the establishment of a separate muftiate for Tatarstan. In August 1992 Gabdul Galiulli was chosen as the new mufti for Tatarstan in

Uzbeks at a mosque prior to their pilgrimage to Mecca

rivalry to Talgat Tadzhuddin, who continued in office in Ufa and maintained his claim to be the spiritual leader for all the Muslims in European Russia and Siberia.³ Although Talgat's official residence is in Bashkortostan, he himself is a Tatar and was educated in Tatarstan.

Local observers think that the split developed largely as an expression of political manoeuvring and power struggles among the more religious Tatars but ostensibly the cause of the split was the depiction the crescent moon, the cross and the Star of David, all together, on the upper windows of a new mosque in Naberezhnye-Chelny. Talgat had seen such symbols on mosques in Turkey and wanted to incorporate them into this new mosque as symbols of tolerance and respect for different religious traditions.

However, Talgat's opponents saw in this their opportunity to set up a rival political (or religious) organisation. They protested against the use of the cross and the Star of David, wanting only the crescent moon — which is widely regarded as a symbol of Islam. Even though the offending symbols

Free market religion

were removed, the pretext was sufficient for a new Mufti to be elected.[4]

The Muftiate entails many political, economic and even commercial factors because large sums of money had been passing through the hands of the Mufti and of the Spiritual Directorate, including money from Islamic states such as Saudi Arabia. The Mufti in Ufa gave grants to enable certain pious Muslims to go on the hajj — the pilgrimage to Mecca and Medina. Nowadays the pilgrim has to find half of the cost himself and the mosques might supply the other half from grants they have received from Saudi Arabia and elsewhere.

The headquarters in Ufa of the Muslim Spiritual Directorate for European Russia and Siberia

Obviously the ability to allocate such grants confers power within the Islamic community. Recently there have been reports that the economic resources available to the Muslim clergy have been used to induce Tatars to observe the fast of Ramadan. Those keeping the fast are said to have each received walnuts and three kilos of raisins from the mosque.

Within Tatarstan itself, some local mosques supported Talgat Tadzhuddin and others supported Gabdul Galiulli but their loyalties were not necessarily constant, changing according to whichever mufti helps them. Both muftis were recognised by the government. Yuri Mikhailov, a Russian who works for the office of religious affairs in the Kazan Kremlin, said that they would prefer to deal with one centralised authority rather than with several, but this process is now independent of whatever the government thinks. It is a religious situation and not a state one, so they had to register Mufti Gabdul Galiulli, and the state could not refuse registration when they brought along the proper documents. Therefore both muftis had their own representatives in

Kazan. However, government attitudes towards Galiulli soured when in 1995 he led a group of students in occupying the offices of a local newspaper, demanding that it be returned to its former use as a madressah.[5] This action might have helped Galiulli to raise support among more militant Muslims at a time when his finances seemed to be suffering: by 1996 Galiulli was under investigation relating to claimed financial irregularities.[6] Meanwhile the Spiritual Directorate in Ufa was still refusing to recognise the independent status of Tatarstan's spiritual board of Muslims, so that they continued to clash over the jurisdiction of Tatarstan's 750 mosques. This issue was not settled until 1998, when the Spiritual Directorate in Ufa finally gave official recognition to the Tatarstan muftiate.[7]

However, in Ufa too questions have been asked about the Mercedes and other expensive vehicles parked outside the headquarters of the Spiritual Directorate. Ordinary imams too have benefitted from the economic and political freedoms since the fall of Communism. With the possible exception of young imams in newly-founded mosques, it is claimed that the Muslim clergy in Russia "cannot nowadays be listed among the poorest members of society."[8]

Another internal conflict within the Muslim community has come in the form of the Ahmadi movement. In an interview with me in 1991, Farid Heidar al Salmani of the Marjani mosque in Kazan said that even though the numbers of different Muslim organisations had been mushrooming, the 'official' Muslims in Kazan felt threatened by the Ahmadi sect of Islam, because the Ahmadis have better quality literature — some printed in London and other literature printed locally on a high quality printing machine imported from the West.[9]

Farid Heidar al Salmani also admits to feeling threatened by Christianity. He says: "Recently a new process of Christianisation has begun. It is a big problem because the schools are not differentiated by religion but Christians go into the schools and try to influence the children. Muslims cannot give the Qur'an free of charge but Christians give the Bible to children as a free book because many foreign Christian organisations help the Christians here."

Archbishop Anastasi of Kazan and Mari told me that such evangelism has been conducted mainly by Protestant groups, because the Orthodox clergy do not enter schools in an official capacity as priests. However, because the Orthodox church is the most visible one, the Muslims tend to assume that any proselytisation has been done by the Orthodox church.

Attitudes towards foreign missions

It is unfortunate that relationships between the Orthodox church and foreign missions have sometimes been soured by the anti-Orthodox stance of some Protestants, including certain foreigners whose negative attitudes towards the Orthodox church have not always been welcomed by local Russian Christians.[10]

Even though such individuals are not typical of most foreign missions, from an Orthodox perspective it may be difficult to distinguish between the different kinds of Protestants. A few 'negative' experiences can create stereotypes and motivate the Orthodox church to try to reduce foreign influence among Russian Protestants.

Nevertheless, the Orthodox church still remains the largest single denomination in Russia. All the different Protestant churches together can claim a membership of no more than a few per cent of the population, whereas somewhere between 45 per cent and 75 per cent of the Russians identify themselves as in some sense Orthodox, although this no doubt includes many who are Orthodox according to their 'culture' rather than their 'belief'.[11] The remainder either feel religiously unattached, or are 'nominal' atheists.

Despite its numerical advantage, parts of the Orthodox church have also felt increasingly threatened in the last few years by certain Protestant groups on account of the Protestants' potential sponsorship from more affluent churches in the West. Russian Baptists, Pentecostals and others have benefitted from literature and finance for church buildings and the salaries of evangelists. The Orthodox church has also felt threatened by the mass evangelism of Protestant preachers who have held evangelistic campaigns in Moscow and some other cities. Western preachers have the financial resources to pay for advertisements on local television and radio, the printing of advertising posters and leaflets, and the provision of free scriptures and other literature which are handed out at the campaigns themselves. Superficially, it appears that large numbers are responding at these campaigns, but in fact many of them fail to become integrated into the Protestant churches.

Are Protestants a 'threat'?

Six months after Vasili Yevchek began to plant a church in the city of Naberezhnye Chelny, he had a congregation of about 500 people. In February 1993 his brother Peter started a church in the nearby town of Yelabuga, which after one month had a congregation of about sixty to eighty people. In Nizhnyekamsk, another city in that region of Eastern Tatarstan, their colleague Stepan Borisov had about 100 people in his church just three months after he had begun his missionary work there. During the same period another 100 people joined the church started by Stepan's brother, Vasili Borisov, in the nearby town of Zainsk. Another colleague, Nikolai Jubak, had a congregation of fifty people two months after he began to plant a church in the small town of Kamskiye Polyani.

On the surface this appears to be a phenomenal growth rate among these new Pentecostal churches in the valley of the Kama river in Eastern Tatarstan. Statistics like this could make some of the Orthodox clergy feel that the Protestants are a 'threat', but this reaction is based on very superficial and limited observations.

A Pentecostal congregation meeting in an apartment in Anadyr, Chukotka

What tends to happen is that there is an initial period of apparent 'success' and growth, but the rate of growth usually slows down after some months and the momentum is often lost. Rapid growth needs to be followed by a period of consolidation and a greater focus on the teaching and training of those who have come to faith. At the same time there is likely to be a certain amount of loss: as in Christ's parable of the sower, there will inevitably be those whose initial enthusiasm is strangled by the cares of this world, or whose roots were not put down deeply enough to withstand pressures or persecution.

Compared to the populations of these settlements, the numbers of Protestants is still insignificant. The wider context of the examples just cited means that 500 Pentecostals in Naberezhnye Chelny still constitute less than 0.1 per cent of the population of 508,163 people.[12] The same is true of the other cities: the Pentecostals constituted about 0.5 per cent of the population of Nizhnyekamsk (numbering 192,525), no more than 0.15 per cent of the 54,360 population of Yelabuga, less than 0.3 per cent of the 36,932 people in Zainsk and about 0.3 per cent of the approximately 15,000 in the settlement of Kamskiye Polyani.

It might be argued, however, that the figures for one denomination do not represent any kind of 'threat' to the religious establishment, but collectively the different Protestant denominations together represent a substantial number. I do not have exact figures for Eastern Tatarstan but to the best of my knowledge the Baptist and charismatic churches are of a similar size, sometimes smaller, than the Pentecostal churches detailed above. However,

Free market religion

while studying the Tatar language at the Academy of Sciences in Kazan in 1991 I also conducted some research on the overall situation in the city — not only among Christian churches but also regarding mosques and the Baha'i religion. What I discovered about the churches was as follows.

The major Christian denomination is of course the Orthodox church, but it is difficult to assess the number of 'members' because many of those who regard themselves as Orthodox only attend the church for special festivals. Moreover, the style of service is such that people often come and go during the course of the service, and can move around to some extent on account of the lack of seats, so there is a problem in assessing the numbers of regular attenders. This is easier for Protestant churches in which there is seating provided and in which people feel a greater sense of obligation to stay to the end of the service if possible (although in practice there is also considerable freedom in this).

I met with Archbishop Anastasi and with Father John (who is head priest of the Nikolsky Sobor and is also responsible for the Cathedral of St Peter and Paul, opened in 1989) but our discussions tended to focus on more general issues, such as relationships with the Muslims and the restoration of former church buildings which had been taken over by the state. In terms of attendance at church services, more specific details were given in an interview with Father Pavel, who conducts services in Tatar for the 'baptised' Tatars (*kryeshchyoniye* in Russian, more usually known as *kryashyeni*, from the Tatar corruption of the Russian word). Father Pavel reckons the total *kryashyeni* community to be "several dozens of thousands" but only about 150-200 come to his services held at times of special festivals. The regular, committed congregation is about 60 people, all older people.

Some of the more 'committed' Orthodox Christians are those who are involved in a lay movement called the 'Brotherhood'. It consisted of just thirty members at the time of my interview with Oleg Sokolov, who had organised this group in Kazan. Oleg had placed an advertisement in the newspaper *Vechernaya Kazan*, in response to which he received not only positive responses but also some threatening telephone calls. The group was able to register with the Council for Religious Affairs but had encountered other practical problems when they attempted to organise certain events. For instance, Oleg had tried to organise a concert of a choir in an Orthodox church and had approached friends of his about having the concert broadcast on local radio, but had been informally told that it would be impossible: Oleg did not receive a direct refusal but acquaintances at the radio station had told him that those who organise musical programmes were prohibiting the making of programmes featuring folk music or spiritual music. Since my interview with Oleg this situation has changed insofar as there are now commercial stations on which it is possible to hire time if one has the financial resources to do so.

During the Communist period there had been a registered Baptist church

After Atheism

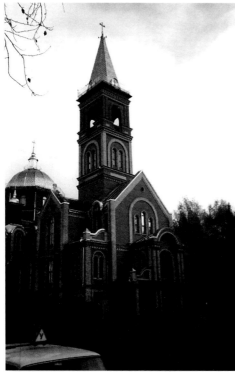

A Baptist church in Syktyvkar, Komi Republic, which has tried to combine Orthodox and Western forms of architecture, and some of its worshippers

in Kazan which I did not manage to visit but I was told that its members, numbering about a hundred or so, were mainly older people who are quite conservative. However, I did visit what for many years had been the unregistered Baptist church. It was meeting in a building on the edge of the city which was packed out with possibly up to a hundred people. One family told me about a KGB raid on one of their services a few years previously in which women had been dragged out of the building by their hair. To a large extent the style of the service was quite traditional — the men in the front rows leading the service while women (all wearing headscarves) and children were relegated to sitting around the walls. Some of the more 'progressive' elements in the church wanted to be less rigid in their attitudes and also to be more open to contacts with other churches (both within Russia and abroad), so a few years later the majority of the members split off to form a separate congregation. I estimated the attendance at the new congregation to be around two hundred in 1997 — indicating an approximate doubling in size over five or six years.

The 'Nazareth' church is another

Free market religion

Baptist church in the city which in 1991 had recently been established by missionaries from the 'Light of the Gospel' Mission based at Rovno in the Ukraine. Later, after the dissolution of the USSR into independent states, the Volga-Ural branch of the mission set itself up as an independent entity. On my first visit to the church in 1991, a 23-year-old young man who led some of the meeting told us that when he first came in May 1991 he was the only one of his age, but then a nucleus of younger people formed as he brought his friends. At first the congregation consisted of mainly older people but the average age was becoming younger and some of the ideas of the younger Christians were also becoming more accepted. He reckoned that the regular attenders numbered about 70 to 80 people, but there are also casual visitors who come to see what it is like but may not come regularly. I estimated the attendance six years later, in 1997, as about two hundred. This again indicated an approximate doubling during half a decade.

My visit to the Pentecostal church happened to be on a Sunday when the pastor and main leaders were absent at a funeral, so I was unable to find out much about the history or membership of the church. Those present in the congregation at the time numbered probably in the region of two hundred.

However, a flourishing charismatic church had been founded by a young couple named Sergei and Julia Borisenkov, who had grown up in the Pentecostal church. At that time the only Protestant churches in Kazan were the Registered Baptist, Unregistered Baptist and Pentecostal churches, but, according to Julia, most of these Christians were older people who had become quite inward-looking and conservative after years of trying to protect their faith under a repressive regime. In 1988, with the advent of more tolerant governmental attitudes towards religion, Sergei and Julia felt it was time for the church to start being more open about their faith too. They started off going to the one place in the city where people were asking questions about spiritual issues — a club called the 'spiritual society' which was largely interested in occult and 'New Age'-type issues. However, a number of these people had found that they were not satisfied with occultism etc. and began to come round to the home of Sergei and Julia to discuss matters further and ask more questions. Julia said that it seemed as if virtually all day they had a stream of visitors to their home, many of whom became Christians. In the end there were about sixty people who wanted to be baptised and take communion, but there was no suitable church for them to join. The older denominations, including the 'Old Pentecostals', still had many old-fashioned practices and attitudes (regarding hymns, headcoverings, forbidding make-up for women, and so on) which were quite inappropriate for these new converts. Sergei and Julia had not set out with any intention of forming a new church, but in 1989 they found that this was the only course open to them. The Pentecostal church gave their support to this new venture.

On my first visit to this new church in 1991, there were 79 committed members but the average attendance was between 100 and 200. Most of them were younger people: about ten to fifteen people were in the thirty to forty age range, another fifteen older than forty, and the rest were students or young people. Over the next few years the church grew rapidly, and by 1995 had already become the largest Protestant church in Tatarstan, having about 700 members, plus another 100 or more attenders who were not yet full members. They had also started church plants in Naberezhnye-Chelny and Novokuibyshevsk.[13]

This 'Cornerstone' church — the name by which it had been registered —

Charismatic churches (like this one in Ufa, Bashkortostan) are among the more rapidly growing churches, attracting many young people

had by 1995 a salaried staff of six people. Sergei Borisenkov continued to be the founding pastor but he was also training others to take on leadership roles. Many in the church seemed to appreciate the preaching skills of a young man named Roman, who later became the leading pastor. At first Sergei devoted himself to administrative matters but after a while Sergei and Julia both became unhappy at some of the emphases in Roman's style of preaching. To some extent he appeared to be modelling himself on some of the American charismatic evangelists like Benny Hinn whose videos they had seen, but there also seemed to be developing a kind of 'personality cult' in which others in the church appeared to be reproducing Roman's mannerisms, vocabulary and theological emphases. Finally the Borisenkovs felt that they had to leave the church they had founded and start again. By 1998 their new congregation had about two hundred people in attendance.

Free market religion

Another charismatic church was founded in the summer of 1991 after a mission organised by the 'Calvary Chapel' denomination in the USA. Initially this new church called itself 'Mission Golgotha', being based on the name 'Calvary', but later they simply called themselves the Church of Christ. Their pastor, Alexander Polishchuk, told me that at the beginning "many" — perhaps 100 or so — became Christians when they came to hear the American evangelist, but only about thirty remained — although some went to other churches. After that the church grew to some extent: on my first visit in 1991 I estimated the congregation to be about fifty-five to sixty people. Since then there has been some growth, but the church then suffered from restrictions introduced in November 1993 by the mayor of Kazan regarding the use of public buildings for religious purposes. For a while the church was meeting mainly as house groups rather than as a combined congregation. They now have a place to meet but I do not have current attendance figures.

In Kazan there is also a Seventh Day Adventist church. Their pastor reported in 1991 that a small group of less than thirty people had now grown to more than fifty. I do not know their current membership but would not expect their growth rate to be too dissimilar from that of the Baptists, in so far as the general style of worship is closer to that of the Baptists than of the charismatic churches.

There have also been a few other groups but they have generally been small and sometimes rather marginal with some unorthodox ideas. One of them which I visited in 1991 was meeting in a small room belonging to a Palace of Culture. There were three tables but all the eleven men sat round the centre table while the nineteen women sat round the edges. Various men gave sermons or Bible readings, or led in prayer. The women all wore head coverings, at least for prayer. During one sermon by the man who had founded this group, what was depicted as the 'Communist' view of male-female relations was contrasted with what he regarded as the so-called 'Christian' view: this he depicted as a hierarchy with God at the top, then Jesus, then men, then women. At the end of the service all the women got up and left, while the men remained for a business meeting. I was later told that this group's founder had previously been thrown out of the Baptist church on account of his homosexual activities. A year or two later he was murdered, apparently because of business dealings with the mafia which had turned sour.

Another 'Christian cult' deserves particular mention because of the way in which it gained a bad reputation in the city. In November 1991 an American evangelist named Bob Weiner had conducted an evangelistic campaign in Kazan, and had asked Sergei and Julia Borisenkov to help with the publicity and other aspects of the campaign. However, Weiner then insisted on starting his own congregation with the new converts. Rather than staying in the city to teach them himself, Weiner appointed as the new pastor a student in his late teens named Nikolai Solomonov. Gradually the new congregation — calling themselves the 'Spirit of Life Church' — began to acquire a reputation for being almost like a cult in the way in which it seemed to put excessive pressures on its members, most of whom were teenagers and students. Some of their parents began to accuse the church of brainwashing their children and even of encouraging the children to steal valuables from their parents in order to help the church's finances. A police investigation was launched and Solomonov went into hiding.

In the meantime, however, the damage had been done. Non-Christian Tatar friends of mine told me about this cult which had been making a nuisance of themselves in public places. For instance, one lady said: "I was with my son on a tram when this bunch of teenagers came on and started shouting at everybody that their religion was the only right one and that my religion (Islam) was wrong. They said we'd go to hell. They shouldn't shout and disturb people like that on public transport, and I certainly wouldn't want to have anything to do with that kind of religion." Another Tatar, belonging to the Baha'i faith, also criticised the Spirit of Life church for its intolerant attitude and the abuse of public transport for proselytising purposes.

Yuri Mikhailov, a government official at the Council for Religious Affairs in Kazan, told me that the restrictions introduced in November 1993 were directed particularly against this Spirit of Life Church. From his perspective, one of the main problems with Solomonov's group was the fact that the Council for Religious Affairs was receiving complaints from parents of converts aged less than eighteen years. "Over the age of eighteen," explained Yuri, "people can be citizens of the State and can become pastors or religious officials, but until the age of eighteen they can only attend church with the consent of their parents. Solomonov, the pastor of this church, stopped studying at the university and did not work. He wanted to start a new church with people of his own age and he told people to stop studying at university — this is bad. By contrast, other pastors encourage young people to study well.

"A second complaint against the Spirit of Life Church was that children were now leaving home and stopped having contact with their parents. Parents came here, to our office, and cried. Some went to the procurator and asked, 'Where is my son or daughter?' Some had gone as far as Moscow. So we took action and refused to allow them to use the Palaces of Culture. I won't communicate with Solomonov and won't move from my position, even though we've had about thirty journalists here asking questions about the situation.

"Another cause of complaint was that Tatars also went to this church. They attracted young Tatar boys and girls and some Tatar parents didn't like this: they wanted their children to go to the mosque and didn't want to allow Solomonov to take away their children if the children were not of age to decide for themselves. In this the borderline between nationality and religion is very thin.[14]

"Unfortunately," concluded Mikhailov, "Solomonov has compromised other Protestant churches because now many people think that all Protestants are bad. So Solomonov was in fact bad for the Protestants too — even an enemy of the Protestants — because people came to the State and asked us to intervene against them. The press also carried negative reports about the Spirit of Life Church. He compromised the Protestants and caused problems for the others because people asked if other Protestants were like Solomonov. So now many people think badly of all Protestants and don't distinguish between them. Solomonov couldn't understand our point of view and our actions and felt that he was being persecuted for his faith."

In conclusion, it can be seen that the overall number of Protestants remains a tiny fraction of the population of the city. At the time of my survey in 1991 there were about 1,000 Protestants in the Baptist, Pentecostal and charismatic churches combined, plus one congregation of Jehovah's Witnesses whose size I did not ascertain but who apparently managed to fit into one normal-sized meeting hall. However, the population of Kazan in 1989 was 1,087,584 — therefore the Protestants represented in the region of 0.1 per cent of the city's inhabitants. Since then the Baptists have approximately doubled in numbers and attendance at the charismatic churches (especially the Cornerstone church) has multiplied by a factor of perhaps four or five, but the overall number of Protestant Christians remains merely a fraction of one percent of the population. Therefore it appears as if the Orthodox reaction to Protestantism seems to be out of all proportion to the real numbers involved. The small numbers of Protestants can hardly be seen as a 'threat', unless their numbers appear to be magnified in the eyes of the Orthodox establishment. If so, the most likely explanation for this reaction probably lies not in the actual growth figures of the Protestant churches themselves but in the impression conveyed by Western-style mass evangelism.[15]

Problems of evaluation: the 'mission business'

"Last summer 34 American missionaries came to our city. They'd each paid $3,000 for their tickets here, and the whole group took up half of one floor of the hotel, but it was like a pantomime: one preached, through an interpreter, and thirty listened — or made videos of it!! They preached for three days and saw 100 converts — whom they baptised in water, even though they knew hardly anything at all about the real situation of each person. After one month only fifty remained. Then less and less people came to the meetings, so the newly-established church really had a nucleus of only twenty or thirty new Christians."

This is how Alexander Polishchuk in Kazan described the origins of his church. He went on to say how a sister church relationship had been established between his church and the one which sent the missionaries: "They want to come back again this summer, and the whole pantomime will be repeated once again. What they want to do is evangelise — they don't ask me what I want them to do. Most of them just come to see the Russians, have a look at how we live, and then go home again with their photographs and videos. Of course, they want me to organise it all for them, which I will, because I suppose I'm getting used to it now!"

Further east, in the city of Ufa, a Russian Christian who had studied at a Bible college in the West expressed a similar attitude. His impression of most Western missions is that of bureaucrats who sit in comfortable offices in London or elsewhere but do very little in practice to help people in Russia — despite claims to the contrary. Whether or not his impression is accurate — in so far as he later revised it by toning down his criticisms — it reflects the attitudes held by a number of Russian Christians.

A 'novelty' book for many Russians . . .

At the same time, they continue to be grateful for the material help, through literature and other resources, which many Western missions are providing. During the Communist period, Western ministries provided a much-needed lifeline for many struggling churches and in general a mutually beneficial relationship had developed. However, the greater freedoms in contemporary Russia have meant that different kinds of relationships are developing.

These relationships can be both positive and negative. Positive contributions by Westerners include the provision of literature and other forms of material aid, and also an expertise in particular areas where there has previously been a lack of teaching or experience. However, there are also dangers in the relationships, particularly in the desire for Western goods and foreign travel. For instance, a pastor from Russia who had been invited to visit England said that his congregation expected him to bring back a video recorder and other Western goods. Such expectations can easily produce a dependency mentality.

In a city in the central Urals there were suspicions that some of the apparent 'converts' to a newly-established Methodist church were possibly 'rice Christians' who wanted free trips to America. I was not able to verify this myself, but local people described how a group of Russian female 'converts' (whose sincerity was doubted by others) appeared to be very protective of their relationship with the visiting American pastor and seemed to keep him from speaking with those who might make him doubt the motives of these women. At least one of these women was later given an all-expenses-paid trip to visit the pastor's church in the USA.

... the Bible

Similar behaviour was shown in May 1992 by a Pentecostal church in the southern Urals. A Scandinavian mission wanted to establish a sister-church relationship between this church and two Scandinavian churches. However, when visitors from Sweden came to visit this relatively small and struggling church, they were carefully kept from knowing that another charismatic church was conducting a major mission in the same city that very week. As a result of that mission, the congregation of the other charismatic church rose from about 100 to more than 250 people. The Pentecostals feared that their valuable contact with Western churches would be lost to this more lively charismatic church in the same city.

At first one apparent 'advantage' of Western preachers seemed to be their curiosity value. In Omsk a local Baptist told me that many people went along to hear the American evangelists simply for the sake of seeing and hearing the foreigners. However, probably this is no longer the case in Moscow and some other major cities, as indicated by the reactions of some local Christians to Kalevi Lehtinen's 'Mission Urals' in 1994. Prior to the campaign, expectations were raised among local pastors when they were shown a video of the apparently huge numbers who responded at the evangelistic meetings in St Petersburg a few years previously — when both religious meetings and Western evangelists had been a novelty. However, when the Urals mission produced much less dramatic results, the organisers began to wonder if they should focus more on smaller towns where the local people were still unaccustomed to mass evangelistic campaigns.

A problem for some Westerners, however, is a felt need to show results — preferably spectacular ones — to those who have financially supported their missions. Glamorous 'success stories' are often the best way of doing this. A somewhat cynical perspective might say that mission organisations need to produce a constant supply of 'propaganda' in order to justify their financial support. Too often mission strategy can be influenced, perhaps almost unconsciously, by what looks good to supporters. Hence there is a danger of stressing quantity at the expense of quality.

For instance, in 1994 there was a debate in the letter pages of the *Christian Herald* newspaper whereby an organisation called Eurovangelism wished to dissociate itself from being confused with a mission called Eurovision which was advertising an evangelistic campaign in terms which Eurovangelism considered to be exaggerations. For example, Eurovision's description of Siberia as a land with "No God, No Bible, No Hope" seemed incompatible with the fact that there were existing churches in many of the very cities and towns being targeted by Eurovision's mission. I had not intended to let myself become involved in this debate, but I got drawn in when the editor of the *Christian Herald* asked me to try to give an objective appraisal of Eurovision's claims.[16]

In fact, I had little or no quibble with the basic facts as presented by David Hathaway, the director of Eurovision, in the report on his mission which he sent to the *Christian Herald*. The problem lay more in his style of presentation: "Over 50,000 came forward to receive Christ, over 1,000 documented healings — just those prayed for by myself, not including those prayed for by team members — 17 Crusades — 17 different cities — 55 separate meetings — several new churches planted — approx. 30,000km travelled — all in 63 days . . . In every Crusade almost every unbeliever came forward to receive Christ . . . In Susuman 25 per cent of the population came forward to make a decision in two days in the stadium."

These quotes from David Hathaway's report illustrate the way in which it is possible to give an impression to one's sponsors that they had received value for their money. I have little doubt that the figures are generally accurate regarding some of these smaller Siberian towns where Westerners still have curiosity value. However, what does it mean in reality?

I tried to ascertain the opinions of local Christians in some of these Siberian towns. In Yakutsk they laughed when they read Hathaway's report — one of them commenting that the whole population of Russia must be already Christian if one adds together all the numbers of converts claimed by various Western missions in Russia! The fact is that very often Russians have gone forward in response to evangelistic messages because they feel that the evangelist expects them to do so, or else they feel that is a "religious" thing to do — rather like the new Russian fashion for wearing a cross around one's neck. Even if they went forward, they have not necessarily become Christians, and in fact relatively few actually end up as church members.[17]

Free market religion

Another Christian wondered why Hathaway should be so specific about the numbers of converts, crusades and meetings but so vague about the exact number of the "several" new churches which were said to have been planted. A possible reason for this vagueness was provided by a telephone interview which I conducted in Russian with Iida Lukina, a Christian in the southern Siberian city of Neryungri. She said: "Hathaway's people wanted to organise a new church in our city but they met with the pastors and decided that they did not need to do so because our lively charismatic church was already in this city." However, Iida also said: "Many people repented in the meetings, but no one new has come along to our church as a result of this mission." She thought that some of those who responded might have started to attend the local Orthodox church, as most Russians are unfamiliar with Protestant churches. Even though Iida considered Hathaway's mission to have been "excellent" and "well-organised", her comments about the lack of incorporation of supposed "converts" into the local church (or at least the Protestant one) are consistent with the patterns reported by Russian pastors elsewhere after other evangelistic campaigns by Western missions in recent years. Often there have indeed been some lasting conversions, but these form a relatively small proportion of the apparently large numbers who respond at the public rallies; an informed estimate for Moscow reckons that a total of about 2.3 million people in the city are recorded as having made a response at large evangelistic rallies in recent years but membership of the Protestant churches has increased by only about 50 per cent — from around 6,000 members to 9,000 or 10,000.[18]

However, the publicity reports produced by the evangelists for their sponsors tend to report the apparently large numbers who made some kind of initial response. Such 'propaganda' (or 'hype') can also mislead observers from the Orthodox and Muslim faiths who have begun to regard Protestantism as a threat. In order to protect the integrity of their traditions and their symbols of ethnic identity, they have appealed to legislators for state protection. The Protestants have reacted by viewing such restrictions as forms of persecution or as infringements of human rights. In so far as Hathaway's magazine *Prophetic Vision* often refers to prophecies that the time is short for such evangelistic campaigns to be held, it would be an irony if to a certain extent such prophecies might be partially self-fulfilling.

These kinds of Western missions are the ones which are most likely to catch the eye of the general public and of the Orthodox hierarchy, who might be inclined to view the Protestants as potentially 'stealing their sheep'. It is unfortunate that they do not notice the quiet work of various Western Christians living in Russia who have a desire to be 'servants' of the church, whether it is Orthodox or Protestant. Some of them have deliberately decided to join in with their local Orthodox churches in order to help and encourage the local Orthodox priesthood. Others have been involved in charitable activities such as helping children living on the streets of Moscow because they are motivated by a genuine desire to alleviate human suffering. These kinds of inconspicuous but very positive forms of help rarely catch the attention of the public but at a local grassroots level they have been building forms of co-operation between Western Protestant and Russian Orthodox Christians based on practical deeds rather than discussions about theology. In other cases known to me, Western and Russian churches have entered into the religious equivalent of 'twinning' between towns: that is, members from both sides have visited the other church and have each helped or encouraged the other in some way. Sometimes this has resulted in help with practical material needs but probably more important has been the trans-national 'cross-fertilisation' which comes from the sharing of experiences and insights with one another.

All these and many other kinds of relationships have been developed in an inconspicuous manner by those who are not 'advertising their wares' in the religious free market but are instead investing for the future: their investment is not merely financial but often involves a more demanding investment of one's time, energy and talents. This is not the mentality of a foreign competitor out to make short-term profits but rather that of a long-term investor wanting to stimulate the local economy through joint ventures and partnerships. However, the religious market is far wider than merely that of an Orthodox quasi-monopoly having to come to terms with the presence of relatively similar 'wares' being imported from the West, because many apparently 'new' spiritual imports from the East are crowding into the market too. These will be the focus of our attention in chapter nine.

9
Human rights

On Monday evening, 18th May 1992, I arrived in the city of Yekaterinburg after a 27-hour train journey eastwards from Moscow. This was the third successive year in which I was a guest lecturer at the university there, but on my arrival I was told that I would be lecturing later in the week because they thought I would probably like to attend an academic conference on religion which was being held over the next couple of days. Boris Bagirov, the dean of the Faculty of Philosophy at the university, said that I did not need to pay anything for the conference as it was being sponsored by "the Church." Both he and I assumed this was the Orthodox one.

I was therefore puzzled the following morning to see none of my friends from the local Orthodox, Baptist or Pentecostal churches. Soon I learned that the conference was in fact sponsored by an organisation called 'New Era'. They had paid for the participants to stay at an exclusive leisure complex in the hills and forests to the north of the city and had even hired security guards to ensure that the foreigners were kept safe in this secluded spot — and who even followed us around when we went from one building to another or for a walk in the woods!

The conference was entitled 'Encounter of Religions and Cultures: Past, Present, Future'. Among those present was a representative of New Era, who said their organisation was part of the "International Religious Foundation" and was associated with the "Assembly of the World's Religions." She encouraged the participants to take samples of the International Religious Foundation's newsletter and of the glossy brochures about the Assembly of the World's Religions. Paul Mojzes, an American professor who had also been invited to the conference, later explained to me that both New Era and the International Religious Foundation were among the many 'front' organisations for Rev. Sun Myung Moon's Unification Church — the so-called 'Moonies.'

Moon turned out to be the founder of the Assembly of the World's Religions which his New Era organisation was promoting in Russia. Their literature showed that it aims to be a bridge not only "between Europe and Asia" but also between Africa, the Americas and Eurasia. One of their publications described the "Welcoming Ceremony" of their second Assembly, held in San Francisco from the 15th to 21st of August 1990, to which had been invited participants from many different religious traditions, including Buddhism, Christianity, the native American Coastanoan religion,

Confucianism, Hinduism, Islam, Jainism, Judaism, Shinto, Sikhism, the traditional religion of the Yoruba people of West Africa and Zoroastrianism. Participants at this welcoming ceremony had been "invited only after careful review of their qualifications and ability to contribute to the overall purpose of the Assembly," and included Metropolitan Filaret of Minsk and Grodno, the patriarchal exarch of all Byelorussia. Presumably the mention of Metropolitan Filaret conveyed to a Russian audience a feeling that the 'Unification Church' was in some way recognised or accepted by the Orthodox church.

The conference which I ended up attending in Yekaterinburg was simply an academic one, with no elements of religious worship or devotion of any kind. One of the Russian participants commented at the end of the conference that she was impressed by New Era precisely because they did not attempt to convert her. Another participant, who belonged to the nominally Buddhist Buryat people of Siberia, decided that New Era or its parent body the International Religious Foundation might be able to help him in promoting the local cultural and religious life of the peoples around Lake Baikal. He wrote to Thomas Walsh, the executive director of the International Religious Foundation, to ask if they would like to establish a branch of their Foundation in Ulan Ude, the capital of Buryatia.

It was clear that the Unification Church is beginning to make inroads into the CIS under the guise of New Era and the International Religious Foundation. Initially it seemed to be a low-key policy of 'infiltration' and trying to build up goodwill. However, by March 1993 there was a more fully-fledged centre of the Unification Church in Yekaterinburg, which had already extended its influence as far as Ufa, where I met a Bashkort young man who had been reading their literature. The following week I learned that in Kazan a young man in the congregation of Alexander Polishchuk's charismatic church had been espousing Rev. Moon's 'Divine Principles', as if trying to win converts from other churches.

By September 1995 representatives of the Unification Church in Yekaterinburg had already visited many of the city's schools, where their doctrines had apparently begun to have more influence among both staff and pupils than the teachings of any other religious group. This had become a cause for concern within the department of religious affairs in Yekaterinburg's city council. Tanya Tagieva, who works in that department, told me that the authorities had little or no information about these new religious movements except for the Unification Church's own propaganda. The government was also concerned about the teachings being promoted by a so-called 'Hubbard College' and the widespread dissemination of books by Ron Hubbard, the founder of Scientology. Their shortage of reliable information meant that it was difficult to distinguish 'healthy' and acceptable religious movements from 'dangerous' cults and sects which were taking advantage of the increased freedom of religion in Russia.

This suspicion of certain forms of religion was reinforced by some academic papers about the negative psychological effects of certain cults, and of some aspects of occultism. Further fears were generated by reports in the media about the activities of extreme religious cults elsewhere, either in Russia or the West. For instance, in the Spring of 1995 a formerly obscure Japanese religious cult became world headline news when it was widely considered to have been responsible for the release of poison gas among travellers on the Tokyo subway system. Reports circulated about the manner in which adherents of AUM Shinrikyō were brainwashed, malnourished and (among other ritual practices) were induced to drink the blood of the cult's founder, Asahara Shōkō, in order to obtain so-called spiritual 'power'.[1] In Russia, the authorities were particularly concerned about this case because just one week prior to the Tokyo sarin gas attack, a legal case against AUM Shinrikyō had opened in the Ostankino district court of northern Moscow. The cult had about thirty thousand members in Russia, as compared to an estimated eight to ten thousand adherents in Japan itself. Already they had bought a helicopter in Russia, and some Russian journalists speculated whether or not they had set themselves up in Russia partly in order to buy nerve gas on the Russian black market.[2]

Former Communist officials who had previously been taught to have an antagonistic, or at least cautious, attitude towards religion were beginning to feel that in some ways their fears were being confirmed. They had been used to State restrictions on religion, and legislation seemed the most obvious way to restrict the activities of such cults. Probably this was an influence on Victor Pavlevich Smirnov, who during the Communist period had been in charge of the department for religious affairs in the Sverdlovsk city council and in 1996 was the deputy of the regional assembly who tried to push through legislation severely curbing religious activities in the region. Precedents for restricting religious groups had already been set in several parts of Russia, such as the ban on the use of public buildings for religious purposes in the cities of Kazan and Naberezhnye Chelny. In the spring of 1996 the regional authorities in the Sverdlovsk oblast (which includes Yekaterinburg) and in Udmurtia attempted to introduce similar forms of legislation but opposition from existing religious organisations delayed the passing of the bills for several months. Nevertheless, amended forms of the legislation were passed that autumn. In Udmurtia, however, an appeal was lodged with the regional procurator, who eventually ruled that the laws did indeed violate constitutional guarantees of freedom of religion. The regional government backed down and repealed its law.

Meanwhile, pending the procurator's decision, several churches had been forced to find new meeting places. For instance, the 'Philadelphia' church — a leading evangelical fellowship in Izhevsk — had to move from the premises they had formerly hired into a temporary meeting place in a hall seating only 600 people, which was insufficient for their average congregation. Ironically, the new laws initially caused some non-Christians to visit the churches out of curiosity, to find out why they had been banned!

Pentecostals at a baptismal service in the Kama river, southern Udmurtia

The Sverdlovsk regional legislation required the registration of all "missionaries" — a term which even included the local Orthodox archbishop! — along with documents on the nature of their organisation, a "description of their rituals and ceremonies and other religious actions," a copy of their "holy book or code of laws" and other such details. Among the specific restrictions placed on missionary activity within the region is one which forbids mission work among children (particularly those in state schools, orphanages or kindergartens) without their own consent and also the written permission of their parents or legal guardians.

The Sverdlovsk regional department of justice was entitled to refuse registration of missionaries on several grounds relating to unacceptable kinds of activities by the mission. Several of these reasons seem to have been particularly motivated by a fear of the anti-social and psychologically damaging effects of certain cults: for instance, a mission could be banned if it damages the "physical, mental and moral health of people," or if it led to the break-up of families or the erosion of family unity. Other reasons for banning missions included their stirring up of ethnic or religious discord, use of drugs, the propagation of violence or the use of "unlawful compulsion."

The Baha'is

Many other religious movements have also been trying to make converts in Russia and other states of the former USSR. For instance, saffron-robed followers of Krishna can be found in many cities, having grown from eight communities in 1990 to more than 120 in 1995 — including, in the Urals, nine in Chelyabinsk and six in Yekaterinburg, in southern Russia, eight in Rostov on Don and five in Krasnodar, plus seven communities in Moscow and another seven in the Arctic city of Murmansk.[3] Representatives have taken part in events bringing together people from different religions, such as a seminar which I attended in 1992 at the Kazak State University in Almaty. Although a small minority, their lifestyle and dress brings the followers of Krishna to public attention more than many other groups who are not so conspicuous.[4]

The Baha'i faith has also been spreading throughout Russia.[5] It was first introduced to the territory of the former Russian empire in the 1880s, when a Baha'i community had been established in Ashgabat (Turkmenistan) but in the 1930s it was broken up through persecution.[6] In recent years, however, the religion has again been introduced to the former USSR, this time from the West. For instance, it was through American Baha'is that Professor Daniel Pivovarov of Yekaterinburg became a Baha'i. Currently he teaches courses on religion at the Urals State University. He has organised seminars at which proponents of religions as diverse as Islam and Pentecostal Christianity have spoken about their own religions. When some Canadian Baha'is came in 1991, about 300 people attended their presentation, some of whom then became Baha'is.

In 1989 a 'Peace and Environment Festival of Scandinavian Countries' was held in Murmansk. Participants came from Finland, Sweden and Norway. Some Baha'is brought an exhibition of Baha'i literature and explained about their faith. They were not allowed to bring many books but they sent further literature afterwards by post to those who requested it and supplied their names and addresses. Many of those who became Baha'is in Murmansk first heard about the religion through this festival.

Later, some American Baha'is who had lived in Finland for eight years came to give further training in English to English teachers in Murmansk. When the local teachers mentioned about the books they had received, the Americans said they were Baha'is too and that they knew the woman in Finland who had sent the literature. At first the Americans came for a two-week course, but they were also asked to give an hour's lecture on the Baha'i faith. Two English teachers and the Americans' interpreter in Murmansk have become Baha'is, but others have come to faith through Russian translations of Baha'i literature. A Finnish woman also came to Murmansk to help establish the group.

In December 1991 the Baha'is of Murmansk organised a conference on the "problems of the northern peoples." Their intention was to "give an opportunity" for the native Saami people (whose traditional culture is like that of the Finnish Lapps) to become Baha'is. The Baha'is are aware that the Saami are losing their own culture, and that there are many problems among them caused by disease, poverty, illiteracy and alcoholism. Though the Baha'is hope that their teachings might help the Saami to overcome problems of alcoholism and of inter-ethnic relationships, they are less clear about how their faith could help to preserve or even "enrich" the native culture. Some critics might say that the adoption of any outside religion would further alienate the converts from their traditional culture. Pragmatically, however, it has to be admitted that relatively little has been left of that culture after many years of erosion by Russian influences under both the Tsars and the Communists.

By 1991 there were about thirty Baha'is in Murmansk. Five children aged between five and twelve were also attending Baha'i classes; of these, three were children of Baha'is and the other two were children of people who were "interested" but had not yet become Baha'is.

Guzel Khakimova is a Tatar woman working as an English teacher in Murmansk. When she first read the Baha'i books, she felt that the ideas had been her own since childhood. She commented that Baha'is do not "make" Baha'is but rather "find" them. The Baha'is "pleaded" with Guzel for her to leave the school where she has taught for sixteen years in order to help with a travel firm which the Baha'is of Murmansk were setting up to facilitate visits by other Baha'i teachers from abroad and which would also provide English courses and translation services. Guzel agreed to help with this venture, commenting that "any loss of income is not so important because I'd have more time for Baha'i activities and also for my home and family."

Human rights

A Baha'i meeting in Kazan

In 1991 I spent a month living with Shamil Fattakhov, the leader of the Baha'i 'Local Spiritual Assembly' in Kazan. Shamil is a Tatar whose conversion to the Baha'i religion has prompted some more zealous Muslims to view him as a traitor to Islam. He explained to me how he had become a Baha'i:

"In 1989 a group of Americans came to our city. They had not planned to come to Kazan, but for some reason were sent here; perhaps it was because they were trying to promote peace, and our city has a reputation for violence. These 62 Americans needed an interpreter and also someone to introduce them to those who had made a film about the Kazan gangs. I got the job. After their interview at the TV studio, one of them thanked me and gave me a gift — a book about the Baha'i faith. Then another one came with another book on the same subject, and in a few minutes I had three or four books!

"Looking back now, I can see that I had been looking for a faith for at least two years. For my television programme I had been visiting mosques and churches but I couldn't feel a genuine religious spirit there, and I also saw many things which made a bad impression on me. But I felt that there has to be some faith that is equal, that is trying to unite everyone and is universal. I was so naive that I'd started to put down the main principles of the new faith — just for myself, to prove something to myself. At the same time I was still an atheist, but I still saw some need for a religion based on a belief in God and with a supernatural origin.

After Atheism

"It was about half a year later, in December, when the Baha'is came. When I opened their books I felt they were familiar: then I opened my papers and found things in common. I understood I was looking for this particular faith. This surprised me, but I was still so hard an atheist that I still had problems about a belief in God (because of this country's official ideology for so long).

"Then I started thinking about this faith. I told the Baha'is that I liked these principles but I didn't agree with certain points and didn't see a need for

Shamil Fattakhov, President of the Baha'i Local Spiritual Assembly in Kazan

God. Then I invited two of them to have a short interview with me on my television programme and in the evening invited them to my home. Some of my friends were interested in this faith — but I wasn't. I had nevertheless been impressed by the way they put up with the place they'd been put up in — a Pioneer camp with little heating and no hot water when temperatures outside were minus 25°C — and about half of them were from Hawai'i! Two of them had to wash their hair kneeling under cold water. They had each paid a lot of money for the trip but didn't complain.

"A few of the others who were with these Baha'is declared themselves to be Baha'is too. So when the American Baha'is left, we gathered together a couple of times and talked about God. I liked it but thought I'd not be a religious person myself. However, I'd made a deal with an American lady to make a film about the revival of a church and a mosque in the Tatar republic. She came back in March, when I was again interpreting a lot and had a better knowledge than the others. Later, when I was talking with someone who wanted to ask a question and the foreigner was busy, I found myself explaining it as if I were a Baha'i, as if it were something precious to me. So then I realised that I too was a Baha'i and I accepted God.

"I'd also been wanting to stop drinking, and so this helped me stop, and also helped me to be a better person. I began to define smaller things as sins, and to work to get rid of them from my soul."

In September 1990 the first Local Spiritual Assembly was formed in Kazan, when there were about 17 or 18 Baha'is in the city. By June 1991 the number had grown to about 60, and had almost reached 100 by November 1991, when I attended some of their meetings. Another community has been formed in Naberezhnye Chelny, a city about 230km from Kazan, and there are individual Baha'is in other cities on the Volga such as Cheboksary, Ulyanovsk and Samara. The Kazan Baha'is are responsible for propagating their faith throughout the Volga region, the Ural mountains and Western Siberia. By 1994 there were over 5,000 Baha'is in Russia.[7]

To some extent they are helped also by Western volunteers, known in Baha'i jargon as "pioneers." I met one of these, a German Baha'i named Barbara, who teaches German in Kazan but has plenty of time left over for Baha'i activities. On Monday evenings a discussion group, known as a "deepening", was meeting in the flat which she shared with Dina. Other local Baha'is referred to Barbara as a "pioneer" who was helping with the "Baha'i World Mission" but Barbara did not want to be described as a Baha'i "missionary" on the grounds that the Baha'is have no formal clergy.

Occult and 'New Age' influences

Apart from formally organised religious groups like Krishnaites or Baha'is, there is a great proliferation of groups and individuals advocating teachings which might be broadly described as being of a New Age or occult character. Fortune-telling has a long history in Eastern Europe, but some New Age groups have recently arisen through Western influence. For instance, a woman in Yekaterinburg has been organising regular visits to Britain for groups of teenagers who told me that they had spent one day in London and the rest of their two weeks at Findhorn in Scotland. At Findhorn people are encouraged to seek communion with nature spirits who 'build up' plants and vegetation from 'energies' channelled down by *'devas'* (a Hindu term for a god or good spirit). Findhorn, on the Moray Firth, is alleged to be on a 'cosmic power point' and the community there is well known as a centre for ideas which are generally labelled as 'New Age'.[8]

Other New Age ideas could well be more indigenous to Russia — perhaps even to some extent influenced by pre-Christian paganism and shamanic cults in the region. For instance, the *Soviet Weekly* carried an article in January 1990 describing claims that "energies" in trees can affect the health of human beings. In this case, individuals' responses to trees were measured using a piece of wire.[9]

Similar ideas were expressed to me by a Russian lady named Lena who described in some detail how she uses a ring on a piece of string in order to divine the "energies" of a place. For example, a colleague of hers kept complaining of headaches while sitting at her desk, but not while she was elsewhere in the building. Therefore Lena told her to hold up a ring suspended on a piece of string, which moved in an irregular, zigzag fashion when the colleague held it up near her desk. However, in a different part of the room it moved in a smooth circle when held by this colleague. Lena then suggested that the other woman move her desk to the other side of the room because the "energies" in the first location did not suit this particular person. Now, Lena claims, the colleague no longer complains of headaches.[10]

Lena also told me how she uses the same methods to discover areas in a person's body where there are abnormal "energies", claiming that these are areas where there is some kind of ill health. However, she is unable to offer any cure for the sickness.

The concept of "energies" might even be linked to natural phenomena. Lena claimed that the Urals area is special because it is the meeting place of two continental plates and contains a richer variety of minerals than other mountainous regions formed by the junction of continental and oceanic plates. In the same way, she alleged, the Urals are "special" because many psychics live in the region.[11]

One Russian man, on hearing Lena's views, put his fingers to the side of his head, twisted them round and made a comment which was the Russian equivalent of "nuts"! This man and Lena represent the two ends of a spectrum of attitudes and beliefs in which most people do not know what to believe about such matters. They are curious and might be willing to go along to a lecture or demonstration by someone who claims to be an expert. In the new market economy of Russia and Central Asia, there are many 'entrepreneurs' who have set up in business as psychics, mediums and peddlers of occult powers. On a flight from Moscow to Yekaterinburg I sat next to one such man, Albert V. Ignatenko from Nikolaev, in the Ukraine, who told me that he regularly flies all over Russia to give lectures and demonstrations of his psychic powers. An English version of his business card described him as vice-president of the 'Academy of Nontraditional Scientific Technologies' and as 'President of Association' — although it was not clear to me whether this was the 'Ukrainian Extrasensitive Association' or an 'International Association for the Spiritual Unity of Humanity', both of which were also included on his card.

In traditional Russian society, people who were ill or in need of counsel would often go to a local 'wise woman' rather than to the Orthodox priest. Such individuals are said to be still quite common in rural areas and can also be found in cities. Those consulting them usually bring with them a gift, often of food. Russian manuscripts from the 17th to 19th centuries contain a number of folk cures using leaves or flowers which also contain invocations to spiritual beings. Usually the invocations are to God or to Jesus Christ, but they could also be to Satan. Sometimes the invocations were worn as amulets and at times were regarded as forms of protection against the evil eye.

Perhaps the widespread modern interest in the occult could be seen as a continuation of these folk practices. One well-known contemporary occultist is A. Kashpirovsky, who in the late 1980s was featured on Soviet television. The programmes were soon taken off the air because of protests by both Orthodox priests and lay people, but he continues to travel throughout the CIS conducting public meetings. An account published by a Christian mission — and which I have not yet been able to verify — tells of a woman from Moldova who attended one of Kashpirovsky's meetings and whose symptoms of physical illness did disappear afterwards. At a later date she became a Christian and wanted to be baptised at a Baptist church. As was the custom, she took everything off and put on white garments instead. In doing so, she took off a pendant which she had received from Kashpirovsky but had never dared to open. When church officials opened it up, they found inside a note which read: "The soul of the one who wears this belongs to Satan."[12]

Legislation on religious groups tends not to affect the activities of people like Kashpirovsky or Ignatenko because they are individual 'entrepreneurs' rather than religious organisations. In describing himself as an "academician" of the "Academy of Nontraditional Scientific Technologies," Ignatenko conveys the impression of being in the sphere of education rather than of religion. Clearly the educational system is a key target for those wanting to spread ideas of a kind which are either religious or 'para-religious' (i.e. occult or New Age). At one time a Communist ideology had been propagated through schools, where the teachers, as authority figures, had great influence over immature minds. A problem facing any educator, whether in Russia or elsewhere, is that of inculcating healthy values and ethical standards — which are often based on religious precepts — while protecting children from potentially dangerous occult influences.

Orthodox reactions

In view of the former hostility of the Communists towards religion, it is ironic that now the Communists and the Orthodox church tend to support each other in trying to introduce tighter State legislation on religion. This has been the case at both local and national levels. For instance, in Izhevsk the local Orthodox church in the city had been collecting signatures for a petition asking the government to restrict evangelicals and other religious groups which the Orthodox church regarded as cults. This was prior to the regional government's attempt in 1996 to introduce legislation restricting missionary activity within Udmurtia. (Protestants in Udmurtia who had been praying and fasting while also appealing against the new law regarded as an answer to

Handkerchiefs attached to a bush near a church in Armenia. While churches were closed during the Communist period there was a revival of this practice, originally having pagan roots, as a way for people to express their devotion and prayers at sacred sites

prayer the procurator's decision in their favour, ruling that the legislation was unconstitutional.)

Since October 1990, when the Russian Supreme Soviet had passed a law on Freedom of Conscience guaranteeing greater religious liberties, members of the Orthodox church have wanted to introduce amendments in order to restrict foreign religious groups.[13] However, the Orthodox church itself would have been affected along with other religious groups if a proposed law had been passed on Tuesday, 19th March 1996 in the Sverdlovsk region. In Yekaterinburg, the region's principal city, representatives of all the major religious groups — including Muslims as wells as Orthodox, Catholic and Protestant Christians — had voiced their opinions on the legislation to Tatiana Tagieva, who is responsible for the department of religious affairs within the city's administration. She wrote a letter to the regional parliament warning them that, in the event of the law being enacted, the Christians would take the assembly to court on the grounds that the legislation violated not only the Russian constitution but also international agreements on human rights and freedom of conscience. The regional parliament voted against implementing the proposed laws.

In the Sverdlovsk region, this restrictive legislation had been proposed by Victor Smirnov, who during the Communist period had occupied the post now held by Tatiana Tagieva. It is clear that many former Communists still holding government posts would like to bring back the old system.

Human rights

These local attempts to restrict religious freedom in 1996 were repeated again on a national level in 1997, with the same kinds of forces at work. Again there was a coalition between the Orthodox church and former Communists but this time there was no 'higher' authority within Russia which could overturn any ruling of an unconstitutional nature. In fact, Yeltsin at first refused to sign the proposed legislation on the grounds that it was unconstitutional and violated international conventions on human rights to which Russia was a signatory. In July 1997 he returned the laws to Parliament for revision but virtually the same set of legislation was then passed in September and Yeltsin apparently felt obliged to sign it.

On the surface, the new legislation on religion was purported to be a safeguard against sects such as the AUM Shinrikyō which were deemed to be dangerous. However, many observers felt that it was in effect a return to the days of Communism when the state could decide which religions were acceptable and which were to be outlawed. In particular, the law made distinctions between classes of religious groups whereby full legal recognition with the right to use the term 'Russian' in their title would only be granted to those which had been officially recognised for 50 years — thereby restricting it to those which had officially co-operated with the Communists. 'Second class' religions consist of other religious organisations which have to prove their legitimate existence for fifteen years by re-registering annually with the authorities. A third tier are religious groups, which can perform religious ceremonies in property provided by its own participants but, by implication, appear to be denied the privileges accorded to religious organisations.

When organisations apply for registration, local authorities have the right to refer to state-appointed religious specialists in order to verify if they are to be recognised as organisations with full legal rights. However, article 27.3 states that during the 15-year 'probation' period during which they re-register every year they are not to be accorded various rights which had previously been exercised by many such organisations. The specific rights which they are denied are as follows.[14] Alongside them are listed some of the corresponding rights which are guaranteed by international conventions to which Russia is a signatory:

Russian law:	**International agreements:**
'On Freedom of Conscience and on Religious Associations'	*Concluding Document of the 'Helsinki accords' agreed through the Organisation for Security and Co-operation in Europe, Vienna, 1989*
*During a 15 year period of re-registration, religious organisations do **NOT** have:*	*As a fundamental human right, participating states undertake to:*

the right to create educational institutions. (article 5.3)	'respect the right of everyone to give and receive religious education in the language of his choice . . .' (16.6)
the opportunity to teach religion to children outside the framework of the educational program in state or municipal educational institutions, upon the request of their parents or guardians, with the agreement of the children and by agreement with the appropriate organ of local government (article 5.4)	respect . . . the liberty of parents to ensure the religious and moral education of their children in conformity with their own convictions . . .' (16.7)
the right to have attached to itself a representative body of a foreign religious organization. (article 13.5)	'engage in consultation with religious faiths, institutions and organizations . . .' (16.5)
the right to produce, acquire, export, import and distribute religious literature, printed, audio and video material and other articles of religious significance. (article 17.1)	'allow religious faiths, institutions and organizations to produce, import and disseminate religious publications and materials' (16.10)
the exclusive right to institute enterprises for producing liturgical literature and articles for religious services. (article 17.2)	'respect the right of individual believers and communities of believers to acquire, possess, and use sacred books, religious publications in the language of their choice and other articles and materials related to the practice of religion or belief' (16.9)
the right to create cultural-educational organizations, and other institutions, and to found organs of mass media. (article 18.2)	'favourably consider the educational interest of religious communities to participate in public dialogue, including through the mass media' (16.11)
right to create institutions for	'allow the training of

professional religious education for preparing clergy and religious personnel (article 19)	*religious personnel in appropriate institutions'* (16.8)
right to invite foreign citizens for professional purposes, including preaching and religious activity (article 20.2)	*'allow believers, religious faiths and their representatives, in groups or on an individual basis, to establish and maintain direct personal contacts and communication with each other, in their own and other countries, inter alia, through travel, pilgrimages and participation in assemblies and other religious events. In this context and commensurate with such contacts and events, those concerned will be allowed to acquire, receive and carry with them religious publications and objects related to the practice of their religion or belief.' (section 32)*

Moreover, all the successor states to the Soviet Union — not only Russia but also the Muslim republics of Central Asia and the Caucasus — are also participating states in the Organisation for Security and Co-operation in Europe and have committed themselves to ensuring that religious liberty is respected in their territories. Therefore all these states have reaffirmed their commitment also to the 1990 Copenhagen Concluding Document which states that "everyone will have the right to freedom of thought, conscience and religion. This right includes the freedom to change one's religion or belief and freedom to manifest one's religion or belief, either alone or in community with others, in public or in private, through worship, teaching, practice and observance . . ." (9.4).

This freedom to change one's religion runs directly counter to the assumptions of many ethnic groups and institutionalised religions that one particular religion is the 'national' one for a certain ethnic group. The assumption that there is an automatic identity between ethnicity and religion is one which is being challenged in the contemporary cosmopolitan world in which freedom of thought is highly valued and with it a right to choose one's own religion. Seventy years of atheism have left a spiritual void in which no single religion can claim to have a greater privilege than another. There are those who want to reaffirm their ethnic and spiritual roots, and have a right to do so, but others have inherited an attitude of 'internationalism' and a recognition of the rights of others.

After Atheism

In September 1999 Mohammed Adji of the central mosque in Bishkek, Kyrgyzstan, affirmed this human right to change one's religion. His estimate of the ethnic composition of the approximately 200 men who attend the mosque daily for the lunchtime

A market stall in Udmurtia — will state 'monopolies' in the religious sphere force other religions to 'market' their beliefs outside 'in the cold'?

prayers is that about 70 are Kyrgyz, 50 Uighurs and the remainder are Tatars, Dargins, Russians and others. Five or six Russians regularly come — they are not people with Muslim spouses but had chosen to accept Islam for themselves. Mohammed Adji recognised their conversions to Islam as expressions of the right to choose one's own religion freely, and he also included in this right those Kyrgyz who had become members of non-Islamic religions. Unfortunately, however, this tolerance is not shown by some ordinary Muslims in villages of Central Asia who put pressure on atheists and non-Muslims to conform to Islamic customs.

In other parts of Central Asia there have been attempts to suppress certain religions. In most of these areas Islamic political parties have appeared and the authorities fear conflicts of the kind which they have witnessed already in Afghanistan and Tajikistan — even though the war in Tajikistan, while exacerbated by contacts with Islamic fundamentalists from Afghanistan and Iran, appears to have been primarily a conflict between local political power bases rather than being a 'religious' war.[15] Therefore in January 1992 the Uzbek government introduced legislation to restrict religious activities of a kind which might be seen as 'provocation' by militant Muslims prone to resorting to violence. Further restrictions on religious groups were introduced on 1st May 1998 through legislation increasing from ten to one hundred the minimum number of members required to constitute a religious association. President Islam Karimov and Shoazim Munavvarov, the deputy chairman of the State Committee for Religious Affairs, both made it clear from their remarks that the legislation was intended to curb Islamic extremists and other minority religious movements. In 1999 there were reports of state persecution of Protestants in both Turkmenistan and Uzbekistan but later President Karimov issued a pardon to those arrested on fabricated charges.[16] In

Turkmenistan there is now a requirement that a religious group needs to have at least 500 members before it can obtain official recognition from the state. In such ways the liberties guaranteed by the OSCE accords are being eroded by those who continue to see religion as an expression of national identity and as subject to manipulation by the state.

Far more serious social problems facing these same governments are those which arise from corruption among officials, organised crime (the 'Mafia') and a prevailing attitude among many people that dishonesty, deceit and lies are almost the only way they can survive in a corrupt society. Business and trade depend on trust, which is fostered by honesty. Laws can influence people to some extent but basic ethical standards of honesty and truth are learned better by example than by legislation: anyone can see the hypocrisy in politicians or educators who do not live up to their own moral standards. To a large extent, the moral basis for any economy, as expressed through the values of honesty and truth, is learned through religious teaching and, more importantly, the example of those who sincerely try to practise what they preach. Those positive morals which have to form the basis of any workable economic or political system need to be demonstrated in practice on the factory floor and in the market place, rather than simply preached from pulpits. Those religions which are able to do this will benefit the country far more than those which merely mouth platitudes which are politically correct.

As a state-controlled economy gives way to privatisation and a freedom of competition based on market forces, the religious sphere is still dominated by those who want to maintain their former monopoly positions. During the Communist period the state monopolies could sell inferior quality goods because they had no competition but nowadays the consumers are looking for quality and value. It does not matter very much to Tatars and Uzbeks whether their shoes are from Italy or Russia as long as their quality is sufficient for the intended purpose. Bankrupt businesses and bankrupt religions are those which are unable to meet the demands of the customer. Is it really in the interests of the state to continue to subsidise inefficient monopolies?

Another government monopoly during the Communist period had been the provision of charity: no non-governmental charitable aid organisations were permitted. Nowadays such organisations play a vital role in relieving some kinds of hardships because many people have been suffering on account of the economic changes which have allowed a minority to become wealthy while the majority of ordinary people have suffered either unemployment or else inadequate wages (which are liable to be paid months in arrears). Refugees from conflicts in the Caucasus or Tajikistan have imposed further strains on fragile local economies. It was in fact the failure of the Soviet state to cope with a refugee problem in Armenia that led to the formation of the first non-governmental charity in the former USSR. Moreover, this charity was also religiously based, as will be described further in chapter ten.

10
Compassion

'Forgotten children' of Armenia

"Please can you do whatever you can to make my daughter forget she ever had a mother?" The mother has not been seen since she made this request but now that daughter, who is 17 years old, still keeps asking when her mother will visit her again.

"We can't bring ourselves to kill our child ourselves — please would you do it for us?" was another request put to the staff of the Harberd so-called 'school' number two for mentally handicapped children located just outside Yerevan, in the Masis district of Armenia. Many of the children are also physically disabled. They are often left in their cots all day, with their clothes and bedding usually soaking from their own urine. Under the Communists these children had been treated as if they were animals, and were seriously neglected. Some of them had got into the habit of eating their own clothes or bedding. Previously twenty to forty of the children had died each year, but the new director felt he had done well to bring the death rate down to two in his first year there.

On my first visit to this home, I was frustrated that I could not communicate with the children in Armenian, but I was surprised when one little girl imitated my English greeting of "hello!" I then said something else in English, and she imitated that too. *This girl is not as badly handicapped as it appears*, I thought, *but what she lacks is the stimulus of personal care and attention.*

There are almost 160 children in the home, between four and seventeen years of age. After the age of 17 they go to another home at Geramavan, near Sevan. About half of the children are both mentally and physically handicapped. As far as I could tell on my first visit, there were only two toys among them — Fisher Price Activity Centres which had obviously come from some Western charity. I therefore bought from a shop in Yerevan a selection of suitable toys, as many as I could carry, and returned to the home to spend my last available day before leaving Armenia just playing with these children in an effort to show that they had value as human beings.

The home's director, Sedrak Hovhanissian, was formerly vice-president of an Armenian Christian charity, Gtutiun. Through his efforts, there is now more hope for the children because outside agencies have begun to be aware of the problems. Doctors from the French charity Médécins Sans Frontières now visit the home, and by the time I revisited two years later, in 1995, the charity had helped to pay for extensive rebuilding and refurbishing of the home. There are also regular visits by staff of the Gtutiun charity and of a Finnish mission called Patmos who since December 1992 have employed an Armenian Christian physiotherapist named Albert who visits the home every day and tries to do what he can for the children. Twenty-four of those who are less disabled are now being looked after by some of Mother Teresa's sisters of mercy in Spitak, in the north of Armenia, and two have begun to attend schools for normal children.[1]

On my first visit to this home, in January 1993, Armenia was suffering from the effects of an economic blockade imposed by Azerbaijan with the tacit cooperation of Turkey. Gas supplies had been cut off, so there was no heating apart from a few stoves on which wood was being burned. Despite the much more severe economic situation as compared with that of the Communist period, the changed attitude of the director meant that the mortality rate had been brought down because the children were being treated as having value in themselves.

The 'Brotherhood' movement within the Armenian Apostolic church

Sedrak Hovhanissian belongs to the 'Brotherhood' lay people's movement within the Armenian Apostolic Church, which is the national church of Armenia. This church has forms of liturgy which in many ways are not dissimilar to those of the Orthodox churches but it calls itself the Apostolic church because it is reputed to have been founded by the apostles Thaddaeus and Bartholomew. The Armenians became the first people to adopt Christianity as their national or state religion when King Tiridates was converted in or about AD 301. Apostolic churches dating from the fourth century lie throughout the territory of fourth century Armenia — which was about ten times bigger than the present day country of Armenia — but the most famous of these early churches is the 'Mother Church' in Echmiadzin, about an hour's drive from Yerevan. Next to the Mother Church is the residence of the Catholicos, the primate of the Apostolic church.

The 'Brotherhood' (in Armenian, *Yeghpaizaktsutiun*) is a lay movement within the Apostolic church. Their meetings, in public halls or private homes, consist of prayers and preaching or teaching. The content of the preaching at two of their meetings which I attended in Yerevan could have easily come from a Protestant preacher. However, their members also attend services at the Apostolic church and hold the Brotherhood meetings at times which do not clash with church services.

It is very difficult to obtain statistics on the size of the Brotherhood movement in Armenia. I questioned several of its leaders about this, but received the same answer again and again: they said the Brotherhood does not keep any membership lists and that it is very difficult to give any assessment of

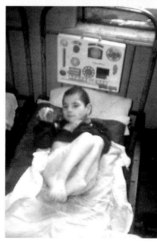

Top: Children at the Harberd school for handicapped children in Armenia.
Far left: A child at the Harberd centre who continues her habit of eating the sheets of her bed.
Left: A child adjacent to, but unable to play with, one of the two activity centres

its size. There are some attenders who are not yet Christians, but also some Christians who do not come to the Brotherhood meetings. When people are converted at Brotherhood meetings, they are encouraged to attend worship services at the Apostolic Church. The Brotherhood simply gather to pray and preach, but the number of their centres has been growing in recent years. They have branches in several parts of Armenia, including Yerevan, Stepanakert, Vanadzor (formerly Kirovakan), Garmul and Stepanavan. They did not have a branch in Gyumri (formerly Leninakan), where prior to the earthquake the inhabitants had a reputation for being very proud of themselves and not wanting to believe in God.

Compassion

I attended two meetings in Yerevan, both of which were held in a public hall ('Palace of Culture'), which is often used for showing films. During the Sunday afternoon meeting there was a power cut, so the only light came through a door at one side and from a candle on the preacher's 'pulpit' so he could read from his Bible. Owing to the serious shortage of fuel on account of the economic blockade by Azerbaijan and the civil war in neighbouring Georgia, electricity was available for only about four to five hours per day and was being rationed to one hour per day in private homes. As the underground railway in Yerevan was usually shut down at 7pm, and hardly any other public transport was available, the Brotherhood had started a Wednesday afternoon meeting at 4pm instead of an evening meeting.

At both the meetings I attended there were about two to three hundred people, but I was told that the attendance was much lower than usual because of the problems of transportation and the sub-zero January temperatures. Under better circumstances the attendance could be two, three or even four times higher. Despite the English translation 'Brotherhood', many of those who attend are women — but at present the leadership is male and the front rows tend to be occupied mainly by men. At certain meetings in private homes many of the women appreciate the opportunities for women to discuss religious issues affecting themselves.

At the large public meetings everyone is seated during the sermon, which is in modern Armenian, but the whole congregation kneels, if space allows, during the times of prayer. After the meeting, friends greet each other and chat. The atmosphere is quite different from the more formal Apostolic church services, during which the congregation remain standing and may find the liturgy difficult to understand because it is in old Armenian.

One of the older children at the Harberd centre demonstrating how she can recite the Lord's prayer

The Brotherhood movement traces its origins to the fifth century AD. It was founded by Mesrop Mashtots, the priest who developed the Armenian alphabet (which is said to have been shown to him in a vision). Although the Brotherhood is close to the Apostolic church and is not a separate denomination, over the centuries its influence has fluctuated considerably. During the Middle Ages (13th-15th centuries) it was weakened through the Turkish invasions and the loss of an Armenian state. Then in the

19th century some massacres of Armenians resulted in the collapse of the Brotherhood. It was revived in 1883 by Archimandrite Voskerchan, an Armenian priest whom the Turks later stoned to death. Under the Communists the Brotherhood was again suppressed but it was secretly reintroduced in 1947 by a group of Armenians from the diaspora in the West who visited what was then Soviet Armenia. On account of the KGB, they could not have official links with the Catholicos but during the independence movement the Brotherhood again became more open in its activities. It is now officially recognised by the Catholicos as a part of the Armenian Apostolic church, which gives their preaching a much greater authenticity in the eyes of most Armenians.[2]

The 'Mother Church' of the Armenian Apostolic church in Echmiadzin

Gtutiun

The first non-governmental charity to be officially allowed to operate in the former Soviet Union was set up in Armenia in 1988 under the name of Gtutiun, meaning 'Compassion'. All its members belong to the Armenian Brotherhood. True to their name, they have been demonstrating the compassion of Jesus Christ by visiting hospitals, children's homes and old people's homes, caring for the sick and lonely, engaging in medical work, providing free medicines to those with doctors' prescriptions, setting up workshops to provide employment for the handicapped and helping refugees to develop new skills. As Christians

Compassion

within the Apostolic church, they also seek to teach people about Jesus through their evangelists and children's workers. The charity has also published a large number of Christian books and has set up a Christian bookshop in the heart of Yerevan.

Gtutiun has had about 300 salaried employees — all of them Christians — in addition to many other volunteers. Three medical doctors and nine pharmacists worked for Gtutiun. The pharmacists dispensed medicine free of charge to those coming with doctors' prescriptions to the six dispensaries established by Gtutiun in different parts of the country. In the areas of northern Armenia affected by the 1988 earthquake, Gtutiun has set up workshops to provide employment for handicapped people who can use sewing and knitting machines. Some of their products are sold at a Gtutiun charity shop in Yerevan which helps to provide revenue for the charity. Refugees are also being retrained as farmers or carpet weavers.

A service at the Mother Church

One reason why Gtutiun does not expand its staff further is their requirement that all the staff should be 'deep believers' and have the ability to speak about Jesus or to preach. A secondary reason is financial. Initially Gtutiun received many contributions from local Armenians as well as from the West, but the deteriorating economic situation in Armenia now means that they are increasingly dependent on Western support.

One aspect of Gtutiun's ministry has consisted of a medical aid group, with three medical doctors and nine pharmacists. Another twelve or thirteen hospital doctors and three or four surgeons also help Gtutiun from time to time by visiting patients in need. Most of these are poor people lacking the money to attend a clinic or with nobody to take care of them. Through foreign sponsors such as the Aznavour Association in France, they have managed to send abroad for medical treatment about twelve or thirteen patients a year, particularly certain children and earthquake victims.

Several pharmaceutical dispensaries have been established by Gtutiun in

Members of the Armenian 'Brotherhood' movement praying during one of their meetings

different parts of Armenia and Karabakh. Gegham Antonian and Suren Apressian, who ran the dispensary in Yerevan, every day supplied medicines to about 500 to 600 people, who came from all over Armenia and some from Karabakh, Georgia and Russia. Theirs was the first charitable pharmacy in the whole of the former USSR. Gtutiun has established other dispensaries in Vanazor (formerly Kirovakan), Gyumri (formerly Leninakan, in the earthquake area), Spitak (a town which was almost completely destroyed by the earthquake), Sisyan, Goris (a town in the south of Armenia which was subjected to severe shelling from the Azerbaijani army) and Karabakh.

These dispensaries provided medicines free of charge for those who have a doctor's prescription. Those without a prescription were examined by Gtutiun's own doctors before any medicines were given out. However, their supplies of medicines are now almost exhausted because they seldom receive fresh supplies. In 1989 (after the earthquake of December 1988) large quantities of medicines were donated by Western charities, and in 1990 smaller amounts were donated from France and Germany, plus a little from the USA and Switzerland. Since then western supplies have almost dried up, partly because Armenia was no longer headline news and partly because of the closure of overland transportation routes during the economic blockade when Turkey was tacitly supporting Azerbaijan and the civil war in Georgia had cut off overland routes from the north.

In the areas of northern Armenia affected by the 1988 earthquake, Gtutiun set up workshops to provide employment for handicapped people. They use sewing and knitting machines which were donated by the Lutheran church in Germany, the Church of the Nazarene and elsewhere. Three centres have been set up — in Vanazor, Gyumri and Spitak. Some of their products are sold at a Gtutiun charity shop in Yerevan. Refugees from Karabakh are also being retrained as farmers or carpet weavers, and a few talented children from poor families are being provided with scholarships for further education.

Gtutiun staff opening a box of medicines supplied by British Christians

Those who work in Gtutiun's visitation department go to people who are poor, lonely or in need. They wash and iron clothes for old people, look after the handicapped, make meals and feed them. These groups visit hospitals, old people's homes, children's homes and the mentally ill. Gtutiun also has two choirs which visit hospitals and old people's homes. One choir specialises in traditional hymns and the other sings modern Christian songs.

Gtutiun has also opened an old people's home in Karabakh and a charitable canteen. Another of their projects is to provide accommodation for orphans who at the age of eighteen can no longer live in an orphanage but have nowhere else to live: owing to the shortage of accommodation throughout the CIS many adults have to live with their parents and may continue to share such a flat after marriage and having children of their own.

Gtutiun's charitable work has also extended outside Armenia and Karabakh. When two passenger trains were blown up outside Ufa, in the Bashkort region of Russia (where leaked gas had accumulated in a hollow alongside the railway line), Gtutiun volunteers were the first outsiders to arrive on the scene to help. They took with them medicines and financial aid. Gtutiun has also sent volunteers to help with earthquake relief in Georgia, Tajikistan and Iran.

Being essentially a Christian charity, Gtutiun also set up a 'preaching department' with full-time evangelists and children's workers. Sometimes they have rented a public hall in the centre of Yerevan where they showed Christian films. From December 1989 to February 1997 Gtutiun also published a weekly newspaper, called *Lusavorich* ('The Enlightener'), which presented, from a Christian perspective, news of current events and items of general interest; one of its most popular sections was a question and answer page on spiritual issues. Its publication came to an end on account of a shortage of funds but since 1997 a Christian sponsor has enabled Gtutiun to publish a 32-page monthly children's Christian magazine called *Avarair* with a circulation of 1,500 copies.

Gtutiun has already published several books, including one entitled *The Covenant of Repentance* by Grigor Tatevasi, a 13th century Armenian priest, which skilfully addresses many theological issues of relevance today. Another book published by Gtutiun is entitled *Let's Know Our Church*. They have also published calendars of scripture readings, a children's scripture colouring book, two textbooks for Sunday schools, a pocket concordance, a song book, a book on the meaning of baptism and translations of some books by foreign authors (including Billy Graham's books *Peace With God* and *Born to Die*, and André Muri's book *Absolute Surrender*).

In 1991 Gtutiun opened a bookshop which is strategically located in the centre of Yerevan, just a few hundred yards from the Square of the Republic (which previously had been called Lenin Square). This was the first and only religious bookshop in Armenia, but in 1992 other branches were opened in Kirovakan, Goris and Stepanakert. At the time of my first visit to the Yerevan shop, the manager, Murad Iskajian, told me that they had 42 titles in stock, all of which are Christian books. Most of them were published by Gtutiun, some of them by other publishers at Gtutiun's request. Murad estimated that they sell more than 50,000 books per year. The Bible is the most popular book, followed by *Peace with God* and George McDonald's *Not Just a Carpenter*. Other popular books are those by medieval Armenian authors such as Narekatsi or Shnorhali which have been translated into modern Armenian. Christian books in English, when available, are also popular, and are used as textbooks by teachers at a nearby school which specialises in English language. The staff also give out large quantities of literature free of charge to genuine enquirers who cannot afford to buy the books. To some extent they have been enabled to do this on account of some of the books having been donated by foreign organisations which gave permission for Gtutiun to sell the books and use the proceeds for other literature projects.

Gegham Antonian and Suren Apressian at the dispensary in Yerevan

On an average day nearly 1,000 customers come into the shop. Some of them are Pentecostals or those from other Protestant denominations who appreciate the fact that, although the staff belong to the Apostolic Church, they preach Christ rather than any one denomination. Many customers, how-

Compassion

ever, are those who had been brought up under atheism and have genuine questions about religion. Some come to ask the staff about passages in the Bible which they do not understand. The staff often spend time speaking with such customers, and often invite such people along to Gtutiun preaching

The Gtutiun bookshop in central Yerevan

centres. Customers with special questions about religious topics can also meet with a priest from the Armenian Apostolic church who comes to the shop on Tuesdays and Fridays. A room at the back is made available for him, and it is often filled by as many as twenty or thirty people who have come to talk with the priest. The most common kinds of questions are those concerning eternal life and what happens after death.

Over 15,000 children have attended the Sunday schools which Gtutiun has established in many parts of Armenia and Karabakh. They were organised in close co-operation with the Brotherhood, but the Apostolic church is also now sponsoring these Sunday schools. Most of them are actually held on a Saturday or Sunday, but there are classes on every day of the week. A minibus, donated by some German Christians, transported the teachers from place to place. Those who taught in more than one place belonged to Gtutiun's salaried staff, but there were also volunteer teachers belonging to the Brotherhood who taught the children in their own localities. Some of the children in the higher classes have become teachers themselves after going through Gtutiun's training course and receiving a certification of competence. Owing to a continuing shortage of qualified teachers, Gtutiun's schools at the peak of their work taught 15,000 children instead of the 25,000 who would probably have attended if there were more staff.[3]

Khachik Stamboltsyan

Khachik Stamboltsyan, Gtutiun's founder, had risked his life to establish this charity. In the summer of 1988 there had been many public demonstrations on behalf of the Armenians in Nagorno-Karabakh who wanted independence from Azerbaijan. Khachik was not one of the 'Karabakh Committee' itself but had close links with them. During the demonstrations in the Opera Square of Yerevan he not only spoke on ecology and national independence but also preached about Christ and explained from the New Testament why they needed to get rid of the Communists. That September Khachik went on hunger strike for three weeks in the centre of Yerevan until the government did something to help the refugees from Karabakh. Finally they allowed Khachik to found a charity to help the refugees.

While he was on hunger strike, a vacancy arose in the Armenian parliament and Khachik's name was put forward as a candidate. The Communists proposed the minister of home affairs, but so many people crossed out his name and wrote in Khachik's that the government declared it an illegal election. They could not refuse to allow Khachik to stand in the repeat election that October, when he was popularly elected as a deputy of the Armenian Supreme Soviet. He was also known to be a Christian among all the atheists.

Khachik told me that he had been educated "blindly," as he puts it, under the Communist regime. He used to argue with his father, who did not like Communism. Then in his teens Khachik read a book called *The American Tragedy* and realised he had been deceived. He calculated that the prisoner described at the end of the book, who had been condemned to death, was eating more than ordinary people in the USSR — Khachik remarked: "We lived worse than a prisoner condemned to death." He therefore decided to do what he could to fight against Communism.

In 1959, at the age of 19, Khachik began to be followed by the KGB. He was a student in the university's faculty of physics but was also a member of a secret dissident movement, which in 1965 took to the streets in a public demonstration to mark the fiftieth anniversary of the Armenian genocide under the Turks. As a result, the government prevented Khachik from taking up the invitations he had received to do research in Japan, Germany or Bulgaria and hindered his postgraduate study in Armenia too.

As a way of fighting the government, Khachik then organised a society for the preservation of historical monuments, many of which were being destroyed by the Communists — especially the Christian ones. After the government closed down their office and banned lectures on this topic, Khachik turned his attention to ecological issues. He succeeded in giving lectures on this in large public halls until 1987, when the ecological movement merged with the Karabakh movement.

Alongside these political activities was a growing awareness of the spiritual side of life. As a student Khachik had been interested in the creation of life and studied a number of different philosophies, including yoga and occult books, but was left dissatisfied by them. He had thought that he could change people by his lectures on art or ecology, but then he began to realise that even science, politics and art cannot change a man on the inside: a person can be good at his or her speciality but be bad on the inside. It was then that he read the New Testament and came to the conclusion that only God could change a person within. As a child, he had considered becoming a lawyer so that he could punish those who do wrong, but now he realised that lawyers

Compassion

themselves are not good but that God is the best lawyer. Through his scientific studies, Khachik had also noticed that there were similar structures in the organisation of atoms, biological cells, and the universe, indicating that the creator was the same, and that what goes on in a cell is so intricately programmed that only God could create such things. These thoughts culminated in 1980, when, at the age of forty, he became a Christian.

From the outset of his Parliamentary career, Khachik encountered opposition from the Communists. At his first appearance in parliament they made fun of his Christian faith. Then, on the 14th of December 1988 — during the turmoil following the earthquake in the north of Armenia — Khachik was arrested and charged with inciting public disorder and causing the loss of 110 million roubles through the strikes. They also tried unsuccessfully to charge him with murder. For one month Khachik was detained in an Armenian prison and was then transferred to Moscow for five months until he was released through international pressure. During his detention various churches and relief organisations sent letters and telegrams to Gorbachev. Amnesty International and Mother Teresa also made representations on his behalf and even the Muslim mullahs from Central Asia sent a letter to Gorbachev asking that he should be released because he was a believer. Those who had been arrested were elected as honorary members of the councils of twelve French cities; Strasbourg elected Khachik, and they were going to send their mayor and a lawyer from Strasbourg to defend him in his trial.

Khachik Stamboltsyan in his former office at the parliament building in Yerevan

Khachik told me how he had led two other prisoners to faith in Christ. One of them came to Khachik at night while the others in the cell were asleep. He said to Khachik: "I don't know why, but I feel I want to tell you all the details of what I did — something I've never told the police or anyone else." Khachik's blood chilled as this murderer confessed to him all the details of how he had violently killed his wife, parents and children — seven people in all. Then the murderer asked: "Can your God forgive or help even me?" Khachik replied: "Yes, if you ask him to forgive you."

The murderer wanted to pray with Khachik, but felt he could do so only when the two of them were alone together. Even though it was against the rules of the prison for two prisoners to be left alone together in a cell, Khachik believes that it was God who arranged for precisely that to happen. When the other three prisoners were all out of the cell, the murderer fell on his knees and, crying, asked God to forgive him. Khachik prayed with him. Every night when the other prisoners were asleep this man asked Khachik questions about God.

Another prisoner was a thief who would have had a twelve-year sentence if the charges against him were proved. One day, when the thief was called for questioning, the murderer asked Khachik to pray for the thief. Khachik replied: "No, you must be the one to pray for him." The murderer did so. Some time later the thief came in, very happy, and said: "Khachik, you prayed for me: now I have only three years." The murderer cried with joy, knowing that God had heard his prayers.

Khachik continued to correspond with these men after his own release. By then he was so popular that he was allowed to preach in prisons and take Bibles to the prisoners. Prior to his own imprisonment Khachik had often thought about prisoners and wondered how to bring the good news of Jesus to them, so he believed that God had given him an opportunity to do so as a prisoner himself.

However, he himself was unable to have a Bible with him in prison. His wife, Gayane, was also refused permission to give him one. She even offered to pay for the guards to buy a "clean" Bible (without any secret messages inside) on her behalf, but this request too was refused.

Nevertheless, Gayane told me how she was aware of God's peace, especially while she was interviewed on television. She felt that she should talk only about their Christian faith, and at the end of the interview the camera picked out the cross she was wearing around her neck. Afterwards her sister commented on how serene Gayane had been during the interview; Gayane then indicated the picture of Jesus which had been in her line of vision and said: "He was smiling at me."

In 1993 Khachik received a letter from a prisoner with a twenty-two year sentence who had become a Christian through Khachik. He wrote that now he is afraid only of God's judgement, not of the world's judgement. "Worldly judgement is only for the body, but God forgave my sins and I know I'll be saved. I pray every day for you because I heard about God from you while I was in prison and you came to me. I am happy even in prison because I have found my God. Maybe I was put in prison so that I could hear about God."

Refugees in Armenia are still living in homes made from converted oil tanks in December 1998

Harassment of religious and other groups in Armenia

After his release from prison, Khachik was able to re-establish his charity, Gtutiun, which subsequently grew to become the largest and most conspicuous charity in Armenia. Khachik was also reinstated in the new parliament and was appointed as chairman of the governmental Commission on Refugees and Disaster Areas. This was a huge responsibility, as there were 250,000 homeless refugees and another 500,000 who had lost their homes in the earthquake and continued to live in cold, wooden houses, wagons, or other forms of temporary accommodation. Owing to the economic crisis in Armenia, this department receives a low budget and so cannot do very much to alleviate the situation. (Most Western reconstruction projects ceased in about 1991 because the economic blockade meant that they were unable to import building materials. There are now many half-completed buildings standing empty.)

Within the government, Khachik became aware of colleagues who were accepting bribes and making money by abusing their privileges. He would speak to such people on a one-to-one basis, but if they did not turn away from their corruption he would bring the matter before a small group within parliament. Then if the problem persisted he would then expose it publicly. As a result two different officials fled from Armenia and have not yet been brought to trial.

His stand against corruption meant that those who feared his revelations also sought ways to bring accusations against Khachik. One line of attack was to blame him for the lack of electricity during the early 1990s because he had advocated the closure of the atomic power station. In fact, Khachik was in prison when the government chose to shut down the atomic reactor because of fears lest it should be damaged by another earthquake. Khachik claims that the shortage of power in this period was a result of corruption whereby the fuel bought for generating electricity had been sold off for personal profit rather than being used for its intended purpose. The fact that Armenia had enough electricity in 1997 even while the atomic power plant was out of production for four months for repairs lends credence to Khachik's claims that Armenia could supply its own needs for electricity even without the atomic station.

Nevertheless, Khachik continued to have opponents in parliament who tried to find ways to oust him from his position or to discredit him. His political rivals in other parties began to spread slanderous accusations about corruption within Gtutiun, trying to make out that Khachik steals aid, has foreign holidays and owns private houses or shops abroad. However, Gtutiun publishes its accounts annually and Khachik has invited those spreading rumours about him to investigate the accounts for themselves. In fact, none of their rumours have ever been substantiated. His flat in Yerevan, where he has lived for many years, is a very ordinary apartment and is subjected to the same kinds of power cuts and other shortages. I have visited both his home and his office in the parliament building when the economic blockade and the lack of heating in January meant that, like all other Armenians, we had to wear warm clothes indoors and eat simple, mostly uncooked, food by candlelight. Therefore I can vouch for the fact that this member of parliament did not have a luxury apartment! I have no reason at all to doubt his integrity.

However, in 1995 all his good work came under severe pressure from the Armenian government itself, which tried to impose prohibitive charges on aid coming from abroad. Under new regulations, 20 per cent of all imported aid goods were to be taken by the authorities, who also wanted to impose stipulations on the use and destination of the remaining supplies. Moreover, the government's demand for tax from Gtutiun was also back-dated by more than a year, so that it was impossible for the charity to pay an exorbitant tax demand in excess of $53,000 — about three times their total annual income! A fine of $28,000 was also imposed on the charity because of an alleged failure to pay tax on the $3,700 with which it was started in 1988. All this was in spite of the fact that Gtutiun was a registered charity and therefore exempt from taxation.

Compassion

In 1995 official tax inspectors had examined Gtutiun's accounts four times and found no fault in them at all. Nevertheless, it appeared from government references to Gtutiun's staff being 'believers' that the authorities had an attitude that Christians were unable to handle financial affairs and were automatically assumed to be incompetent. The irony is that many Western donors felt that they could trust the Christians to channel help directly to where it is needed, without it being siphoned off on the way by local bureaucrats. Therefore many benefactors preferred to give to aid organisations with religious affiliations such as Gtutiun.[4] Eventually this situation was resolved and Gtutiun again enjoyed a tax-free status but Khachik believes that his charity was saved only because of letters sent to Armenian embassies in Britain and the USA in response to publicity about the situation which forced the government to recognise Gtutiun's charitable status. However, their bank account is still frozen and Khachik thinks that instead of a 'quick death' the government is now trying to starve Gtutiun of resources and to close it down quietly. They have had to curtail many of their former activities but are continuing to do what they can through the help of volunteers and part-time staff. They have to supplement their income from other sources because nobody in Gtutiun is paid more than ten dollars per month (when average incomes for many people are $100 or more) and Khachik himself is also helped financially by gifts from a few of his friends.

Khachik lost his seat in parliament in the 1995 elections but many observers claim that the results were rigged. Protests followed both the 1995 parliamentary elections and the 1996 presidential elections; a report released by the US State department states that "serious breaches of the election law and numerous irregularities in the 1995 and 1996 elections resulted in a lack of public confidence in the integrity of the overall election process."[5]

Khachik believes that the next step in silencing him was an attempt on his life. On 31st January 1996 he went out from his office on to the balcony when suddenly a portion of the concrete floor beneath him collapsed. Khachik fell six metres and seriously injured both of his feet and his back. For two and a half months he was hospitalised while three operations were performed on his feet. For the following few months he had to convalesce at home while unable to walk. The doctors expected him to be in a wheelchair for the rest of his life, but Khachik believes it is a miracle that he can now walk, albeit still with a limp.

Was this incident a genuine 'accident'? It is impossible to say. However, the previous day the balcony had been able to support six people; moreover, the building has three such balconies, all of which were in use at the time of Khachik's accident and were in good condition. Khachik thinks that the balcony was chemically weakened in order to kill him. Suspicions of this kind are almost impossible to prove but they are consistent with a prevailing attitude within Armenia that those who resist corruption or other kinds of injustice are liable to become the victims of what appear to be 'accidents', and

Children from Armenian refugee families outside their converted oil-tank homes

that there is little hope of the perpetrators being brought to trial. Similarly, were the tax inspectors investigating Gtutiun merely doing their job, or were they deliberately sent to find financial irregularities? It is impossible to know for certain what motives lay behind their actions, but these events need to be seen in a wider context.

The escalation of the conflict with Azerbaijan into a 'de facto war' has been a contentious issue within both Armenia and Azerbaijan, partly on account of the absence of an official declaration of a state of war and partly because in both countries there were those who were opposed to any recourse to violence.[6] This was not only a question of pacifism, because it also involved ethical questions about what constitutes a 'just war'. The officially undeclared war apparently meant that the Armenian military sometimes resorted to forced and illegal conscription, as documented by Amnesty International and the US State Department.[7] The Armenian government felt that it needed to justify its annexation of Azerbaijani territory in terms of the perceived threat by Azerbaijan of 'ethnic cleansing' in Karabakh, and reports of massacres and atrocities committed against ethnic Armenians there.[8] Azerbaijan was also conducting a similar 'propaganda war' but its greater population, oil reserves and other resources made the struggle seem to the Armenians like an Armenian David against an Azerbaijani Goliath. Politically isolated and with an economy suffering from the effects of the blockade, it seems as if the Armenian government could not afford to allow internal dissent to erode its support even further. Therefore there was a clampdown on potential critics, whether they were political parties, religious organisations or individuals such as Khachik Stamboltsyan who were known to be critical of corruption within the government.

Another possible influence might have been the fact that Gtutiun was actively helping those suffering in Nagorno Kara-

Christmas for a refugee family in Armenia in their hostel room

bakh — and indeed had initially been established to help refugees from Karabakh — because within Armenian politics a source of ambiguity concerns the relationship between Armenia and Nagorno Karabakh once an independent status was achieved. Many Armenians assumed that this Armenian-populated enclave within Azerbaijan would be incorporated into Armenia itself, thereby expanding the geographical territory of Armenia and becoming a symbol of national pride. What they did not realise was that the prolonged separation between Nagorno Karabakh and Armenia had actually led to the creation of a somewhat different local culture — albeit still 'Armenian' — within Nagorno Karabakh. To a large extent the local population tended now to speak Russian as their first language rather than Armenian because Russian had been the lingua franca of the Soviet period with which they had communicated with the Azerbaijani population and with other nationalities. Although still Armenian in terms of official national identity, formal religious affiliation and certain cultural expressions, they were significantly different from other Armenians in terms of language use and aspects of their general outlook.[9]

Armenian refugee children receive presents from children in England as part of 'Operation Christmas Child', organised by the 'Samaritan's Purse' charity

Therefore organisations like Gtutiun which might have

been perceived as helping an independent Karabakh could potentially antagonise those factions within the government who wanted to incorporate Karabakh into Armenia. During the same period there were also attacks on members of an opposition party named the Armenian Revolutionary Federation (ARF or Hai Heghapokhakan Dashnaktsutyun), otherwise known as the Dashnak Party, which claimed the support of over 40,000 members and had popular support in Nagorno Karabakh as a movement of national liberation.[10] Divisions over Armenia's support for Karabakh intensified in late 1991 after Dashnak supporters were elected to power in Nagorno-Karabakh. Furthermore, the 1991 elections reinforced the connection between radical elements in Nagorno-Karabakh and their supporters in Armenia. At that time it gave the Dashnaks added leverage against the president of Armenia, allowing it to cast doubts on his patriotic commitment to the struggle in Karabakh. In particular, the president's reluctance to recognise Karabakh became the main criticism on the political front by the opposition.[11]

Several opposition parties were not allowed to contest the parliamentary election of 5th July 1995. Hopes that the Dashnak party would be allowed to operate again as the leading opposition party were dashed when the Government of Armenia sought, on 16th August 1995, to extend suspension of its activities for another year. In this political climate those who were regarded as potential sympathisers of political opposition groups were unlikely to receive favourable treatment by the Armenian government.

A report dated 1st May 1995 by Marina Kutzian, senior lecturer at the Sociology Department of Yerevan State University, stated that contact with imprisoned members of the Dashnak Party was forbidden but informally she was informed that "they were beaten so cruelly that they were in the hospital inside the prison." Moreover, "lawyers at the prison were attacked in their office. Two days ago, one of them, the most active one, was beaten by a group of people who drove up to him in the street. I saw his blue, swollen face, his eyes were invisible after that. It became a regular thing in Armenia to destroy undesirable people in that way . . . In all these cases one finds the same handwriting. One or two cars with people in military uniform, at least some of them in that kind of clothing, drive up to a place and attack people. No reaction of the police . . ."[12]

Her view is shared also by a report from the United Nations High Commission for Refugees, which states: "The year 1994 marked a turning point on the domestic political scene in the Republic of Armenia: there were political assassinations, repression of freedom of speech and assembly, illegal conscription and intolerance for political opposition. Added to this, the energy blockade and economic difficulties have provided a warning that further development and protection of human rights and fundamental freedoms are at risk."[13]

Kutzian also writes: "According to my impression and informal information which I was able to collect, a land-keeping military group did this action. This group was organized in Armenia before our national army was established. But after that, this group were not disorganized or included in the regular army. So, on the one hand, it is not official, but on the other had it is well known that they are supervised by officials, especially by minister of defence. I have a very strong impression that this group is now used by officials against their ideological opponents . . ."[14] Her views are shared by Vagen Sarkisyan, a correspondent to a newspaper in Armenia called *Golos Armenyi*, which on 27th April 1995 published an article stating: "It seems that in our society there is a group of absolutely defenceless people, who can be terrorized and beaten constantly . . ."[15]

The factionalism within the ruling coalition at that time meant that those who formed part of the coalition, but who wished to promote policies which were not officially promulgated as part of the government's public policy, were in a position of having power but lacking authority. On the surface an impression was given of conforming to principles of open and democratic government but in practice those who were regarded as opponents could be subjected to threats and violence. The very fact that the threats were anonymous and the violence carried out by unidentified people meant that those in power could pretend that it was nothing to do with them.

Not only political opponents, but also the press, religious groups and other people have been subjected to physical attacks, assassination attempts and raids on their property. About a dozen newspapers publishing views sympathetic to the political opposition were forcibly closed down.[16] In this overall political and social climate, those with a vested interest in keeping Nagorno-Karabakh as part of the Armenian state rather than allowing it to become a separate country are most likely to be those with strong feelings about Armenian national identity. It is not unlikely that their political attitudes are similar to those of 'ultra-nationalists' who equate Armenian identity with formal membership of the Armenian Apostolic church. Even though Gtutiun's members all belong to the Apostolic church, their activities are of a more 'evangelical' nature, so could be regarded by the ultra-nationalists as similar to Protestants and members of other 'new' religions whom they perceived as a further threat to Armenian national identity.

Therefore it is not irrelevant to note that in April 1995 several non-Apostolic religious groups (Protestant churches, Baha'is and Krishna devotees) suffered attacks which appeared to have had the involvement of the police and the connivance of top-level government officials, in so far as leading church members were detained in an attempt to draft them into the Armenian army. Patrick Henderson (a British Christian working in Armenia for an aid agency) reported apparently co-ordinated attacks on the Pentecostal, Baptist, Charismatic and Seventh Day Adventist churches. A Pentecostal meeting was broken up by five men — two in military field uniform and three in civilian clothes — who took away Romik Manoukian, pastor of the Pentecostal church, on the charge that he had ignored his draft papers. (He had never received any such papers.) At a regional draft centre he was kept under surveillance by a soldier with a sub-machine gun and was later taken to the army police barracks. Friends who were making enquiries about his whereabouts visited that same draft centre but were told, in a long interview with the colonel in charge, that Manoukian's name was not on the draft list.[17]

This was by no means an isolated incident. Henderson's report gives many other instances of attacks on Pentecostal, Baptist and Seventh Day Adventist churches, sometimes involving those in military uniform and using rifles. In several cases young men in the congregations were taken away for interrogation and possible conscription. A similar kind of attack occurred on the evening of 26th April 1995 against Armenian members of the Hare Krishna movement, who were visited by men who at first claimed to be friends of the couple's son but then produced documents purporting to show that they were defence ministry officials. They ransacked the home and threatened to send the husband to the front line in Karabakh, where, they said, he would be killed.[18] Within both Karabakh and Armenia there have also been cases of harassment and imprisonment of those who wanted to be conscientious objectors on religious grounds, especially some Jehovah's Witnesses.[19] In a state facing a threat of war and which relies on conscription to maintain its military forces, any religious groups advocating pacifism might be seen by the government or military authorities as a potential threat or a destabilising influence. This perception of religious groups could apply not only to the more overt pacifists (such as the followers of Krishna) but also to those who speak of loving one's enemies and praying for those who persecute them.

General trends

In conclusion, it can be noted that both in Armenia and Russia several common tendencies are apparent:

1) Religious groups have often been motivated to express their faith in practical ways through social action and humanitarian aid. During the Communist period religious groups were forbidden from engaging in charitable activities such as helping orphans or visiting prisoners, but the present economic difficulties have meant that those running such establishments have inadequate funding from the state and are grateful for help from other sources. Although this chapter has focused on Armenia, many cases from Russia could be cited too, such as the 'Living Word' Pentecostal church in Yekaterinburg. In the early 1990s its members opened a small shop building as a distribution point for food, clothes, medicines and other goods which they give to the city's poor. Each week they also visited a state institution for invalids or old people with no relatives or families to look after

them. There the Christians washed, shaved and clothed these people, cut their fingernails and helped them in whatever way they could. They also distributed clothes, food, medicine and other goods to the poor and destitute. The local authorities, whose welfare system is heavily strained by looking after such people, were very grateful for the Christians' important contribution. However, the problem is that the 1997 restrictions on religious organisations which had not yet been registered for 15 years have meant that these kinds of charitable activities are also having to be curtailed, even though few (if any) non-religious people are likely to take on such charitable activities instead. Any further restrictions on the charitable work of religious organisations are likely to exacerbate even more the problems faced by governmental social services departments, prison services and other departments whose inadequate resources are already overstrained.

2) Those most actively involved in charitable activities are often lay people rather than members of the clergy, although some clergy are involved too. Religious organisations able to motivate their members to be involved in social action tend to be those with a deeper sense of belonging to a community which provides encouragement and support for other members. This is less commonly found in the traditional Orthodox churches, apart from groups within them such as the Brotherhood movement. Therefore those religious groups which are best able to mobilise people to help in orphanages, prisons, old peoples' homes and so on are the ones who are most vulnerable to restrictive legislation prohibiting their activities.

3) 'Traditional' religions such as the Orthodox church in Russia or the Apostolic church in Armenia tend to be seen as symbols of national identity. In itself this can be positive, helping to restore a sense of national dignity and to foster a consciousness of a rich historical legacy. However, there is a danger that the human rights of other religious groups will be infringed by the promotion of one 'national' religion. It is a particular danger in areas where prevailing social attitudes and values had been shaped for many years by the political misuse of an ideology (Communism) which was intolerant of those holding other views.

4) In areas where the ruling political party is afraid of losing its hold on power, restrictions on religious activities are sometimes a cover for the suppression of potential political rivals. It appears as if there might have been some expressions of this tendency in Armenia, where the Dashnak party was banned by the government but religious and other agencies who openly spoke out against corruption in the government, or who did not fit with an ultra-nationalist model of Armenian identity, were subject to harassment. This is not too dissimilar to the situation in Uzbekistan, where in January 1992 a law came into effect which bans Christian churches from trying to teach their

religion to Uzbeks. The overt purpose of this law was to avoid any 'provocation' of a kind which could spark off violence by militant Muslims, but it seems to have also been motivated by a fear of such situations being exploited by Islamic political parties. They did not want a civil war of the kind which had already torn apart Tajikistan.[20] Neither did they want a repeat of the kind of inter-ethnic violence which had occurred in June 1989 when Uzbeks had attacked and massacred Meskhetian Turks in the Fergana valley.[21] In an attempt to prevent religious differences from escalating into ethnic or political conflicts, the government of Uzbekistan has felt compelled to introduce legislation on religion. What is not clear at this stage is whether this legislation is actually treating apparent symptoms rather than dealing with the root causes of unrest in Uzbekistan.

5) An equation of religion with ethnic or national identity means that there is often a danger of religious symbols being manipulated for political ends. One result of imperialism has often been that different ethnic groups have become incorporated into a single state within which linguistic and other markers of boundaries between the groups have become broken down through prolonged contact. When descendants of conquered ethnic groups no longer speak their own distinct language, sometimes religion has become the clearest remaining symbol of one's original ethnic identity. To a large extent this happened in Northern Ireland, but there is a similar danger of the same kind of scenario being repeated in parts of Russia where local ethnic groups would like independence but have become a minority in their own homeland.

6) A fallacy common to all attempts to impose state controls on religion, however, is the assumption that human legislation can impose constraints on spiritual powers. If there is a God, or other kinds of spiritual powers, by definition they are entitled to break human laws which are not binding on them. When they do so, the results are sometimes interpreted by human beings as 'miracles'. A few of the many claims concerning recent miracles in the former USSR will be examined in chapter eleven.

11
Miracles

Miracles in Ufa

Nadezhda Savateyeva used to wear thick-lensed glasses but after being anointed with holy oil she can now see an aeroplane in the sky. Her husband Peter was one of the priests ministering alongside Boris Razveyev at the Church of the Birth of Our Lady in Ufa. This church had been witnessing many other miracles of healing connected with the miraculous flowing of holy oil from a few of their icons.[1]

These unusual events all started in the summer of 1992 when a young man named Lev was praying in front of an icon of the Virgin and Child. Suddenly he heard what he believed was the voice of Mary speaking to him from the direction of the icon, saying: "Do not cease your songs of praise to me. My grace will be with you. Build the church quicker than you expect!" After this, the local Orthodox Christians found that they began to receive much more of the money and materials they needed for restoring the church, which previously, under the Communists, had been used as a cinema.

At the beginning of August 1992 what was described as "holy oil" began to drip from the hands of Jesus painted on this icon. Soon four other icons also began to weep oil from their eyes, hands or, in one case, from the gospel held in a hand. This phenomenon continued for three months, during which four receptacles were filled with the flowing oil. It ran down the icons, often coalescing along the bottom of the frames, and was then collected into ceremonial vessels.

Boris Razveyev, the priest in charge of the church, remarked that the oil exuded by the icons had a very distinctive smell, unlike anything which could be identified naturally. When he opened a vial in which some had been preserved, I too could not identify the fragrance. Boris said that the strong fragrance of the oil used to fill the church, but it changed to a different smell during times of prayer.

During services when the oil was used for anointing people, many miraculous healings were reported. One man, for instance, had a large lump on the back of his head which was going to be surgically removed. He was anointed on his forehead and received prayer in the name of Jesus. The swelling then completely disappeared, so that the operation was no longer necessary.

A woman in her forties was a haemophiliac whose nose was described as "paralysed" and was emitting large quantities of blood. The bleeding ceased after she was anointed with oil.

When a man who was affected by a demon was anointed with the holy oil, immediately his throat swelled out conspicuously and his head began to rotate round and round. Then his head stopped turning, the swelling in his throat subsided and he was healed.

Alexander, a friend of Boris's, was anointed with holy oil when he had a high temperature and fever. After returning home he sank into a very deep sleep and found that the illness had disappeared when he awoke.[2]

On one occasion Boris began to anoint people but realised that the available quantity of oil was going to be insufficient for the estimated one thousand people who were present. He nevertheless continued to use what oil he had left in his bowl. As he did so, he discovered that whenever he looked in his bowl there was always enough oil remaining. They believe that the oil was multiplied in a manner similar to that experienced by the widows helped by the prophets Elijah and Elisha.[3]

Icon with oil flowing down it at the Church of the Birth of Our Lady in Ufa (Photograph by Boris Razveyev)

The oil began to dry up after flowing for three months, but by then visitors from other cities were requesting samples of the oil to use in their home churches. Therefore they began to use some of the oil from the candles which were lit in front of the icons which had been weeping, and miraculous effects were reported from this oil too. Boris also claims that when some of the original oil was added to water the water was transformed into oil.

One effect of these events upon the wider community has been a strong growth in the numbers attending the church. When the icons began to weep, many people began to come to church who had previously been unchurched: they were not transferring their allegiance from other congregations. Boris claimed that another effect on the community was a remarkable drop in the local crime rate. Previously, when the building was used as a cinema, the small park next to it was often a scene of drunkenness and crime. That district of the city was known to contain a large number of criminals, but in 1992 the crime rate throughout that whole area fell to a tenth of its previous level. In citing this statistic to me, Boris commented that the police themselves were very surprised at the change which had taken place in the area. Boris sees it as a sign of the spiritual influence which the church has begun to have in its parish, both at a human level and also, in the view of some Christians, at the level of unseen powers.

Do 'miracles' really happen?

Accounts such as this are certainly difficult to believe, not only for those brought up in an atheistic environment but also for people in the West and elsewhere. This is not only on account of a worldview among many educated people which claims to be rationalistic and supposedly 'scientific' — even though many popular beliefs and so-called 'superstitions' abound — but could also to some extent be traced to a scepticism within Protestantism, which tended to dismiss medieval claims of miracles associated with relics of the saints. Holy oil flowing from icons is likely to be seen in the same way. It is not a Roman Catholic or Protestant form of 'miracle' but in the Orthodox tradition is seen as a sign of God's grace.

The supposedly 'miraculous' flow of oil from icons has been reported elsewhere in Russia at various times, including three churches of Udmurtia in 1995.[4] However, there have also been allegations that in some instances the phenomenon has been faked by the construction of tiny channels through which the oil is poured. Therefore it becomes almost impossible to distinguish between genuine and 'counterfeit' miracles without a detailed forensic analysis of the icons, but this can be hindered by the reverence accorded to them by those who have already transformed the icons into objects of piety.

From the eye of faith almost any occurrence could be regarded as 'miraculous' which an outsider could dismiss as merely 'coincidence'. Reports of miracles of provision (receiving something at just the time it was needed) or of protection (when an accident is suddenly averted) often depend on the timing of events rather than their intrinsic nature. For instance, Sergei, a Russian Christian from Saransk (Mordova), believed that his opportunity to buy an apartment was "miraculous": the factory where his mother worked was offering a grant of six million roubles at a time when apartments cost twelve million. His mother, a pensioner, applied for the grant and was third in line but those ahead of her turned down the offer. Nobody expected a pensioner to be able to raise the other six million roubles. On the Friday morning before the deadline they had none of the money but by that evening friends had lent or contributed money totalling five million. The following morning Sergei visited a company to ask for a loan and was surprised to find that the accountant was actually there on a Saturday morning, but to grant the loan required authorisation from the boss, who rarely came in on a Saturday. As Sergei was leaving, he met outside a fellow Christian who turned out to be the boss, and who authorised the loan. Half an hour before his deadline that Saturday Sergei had managed to raise all the necessary money. Similarly, when a Chuvash Christian in Kazan was desperate for money was it merely "by chance" that she felt she had to check her postbox (located in the ground floor entrance hall of their block of flats), even though she rarely received letters, and there she discovered a letter containing money? Such tales could be recounted from numerous places but they all involve a certain degree of subjectivity in the informants' perceptions of what others might choose to label as so-called 'coincidence'.

At first glance it might appear to be simpler to document claimed miracles of healing like those of Nadezhda Savateyeva because such a dramatic improvement in her eyesight could be verified by doctors. In practice, however, any attempt to find medical 'proof' for healing miracles is a minefield in itself, as I discovered when I tried to follow up cases of reportedly 'miraculous' healings among the 1,890 people who filled in questionnaires for me at a Christian healing conference held in Britain in 1986.[5] To investigate reports of 'miraculous' healings, I first had to obtain the patients' written consent to the release of medical information about themselves, but then the degree of co-operation received from medical doctors varied considerably. Very often the results were ambiguous simply because there are too many 'loopholes' in terms of possible alternative explanations for the claimed healing.

Even when doctors are convinced about their diagnoses, for example, it is still possible for sceptics to claim that they must have made a mistake, for the very reason that the subsequent recovery in itself implies that the diagnosis must have been wrong! In some cases a sceptic can claim that the person might have recovered in any case through 'natural' processes of healing, or else the use of medical treatments in addition to religious practices means that there is ambiguity about which factor was the most important. Most problematic of all is the whole area of psychosomatic illness and the 'placebo effect', whereby illnesses which seem to be very real to the person concerned are healed through an alteration in the person's psychological state.

Anthropologists who study non-western medical systems often report genuine cures as a result of traditional therapeutic practices. However, these cures are usually interpreted by the anthropologists by reference to this wide repertoire of available 'explanations' (offered from a Western medical

Miracles

perspective) for the apparently inexplicable healings. Once one also invokes the power of suggestion, the use of altered states of consciousness (as in hypnosis), and perhaps even the concept of telepathy, the range of possible 'rationalisations' becomes almost endless. Such explanations have been used to account for the recovery of a young man in the Magadan region of the Russian Far East who fell from a building and suffered severe internal injuries just two days before his wedding. As a last resort, when the conventional doctors were unable to help him, he was taken to a Chukchi shaman who undressed the man's body and rubbed it with what smelled like blubber oil. He then took a dog, "killed it, ripped out its organs and placed the kidneys, spleen and lungs on the corresponding areas of the man's body, which he then wrapped in a bearskin." This whole process was repeated four more times, after which the shaman "poured some kind of mushroom soup into the man's mouth and drank some himself." The man later recovered.[6]

Traditional shamanic rituals in Siberia often made use of 'altered states of consciousness' (for want of a better term). Perhaps some of the alleged cures could be attributed to suggestion, the use of drugs with therapeutic properties, or other related factors. The indigenous inhabitants of Siberia also recognised certain 'altered states of consciousness' or forms of behaviour as symptoms of mental illness or of demonic attack — including, for instance, behaviour such as howling like a dog and fits of violence.[7]

Nevertheless, there do remain cases of apparently 'inexplicable' healings which are still very difficult, if not impossible, to explain away in conventional medical terms. In this category is likely to be the dramatic restoration of the eyesight of Nadezhda Savateyeva. A number of other cases of very remarkable healings in answer to prayer have been documented by Rex Gardner, a consultant at Sunderland General Hospital, who writes that the word "miraculous" can be used as a "convenient shorthand" for otherwise almost inexplicable healings which occur after prayer to God.[8] Not only his article in the *British Medical Journal* but also his book on healing miracles document many such cases from around the world for which detailed medical evidence is available.[9]

Probably many other such cases might be included too, were it not for a great uncertainty about the extent to which possible psychosomatic factors might influence outcomes. This is why I was particularly impressed by a case I followed up myself in which a cancer disappeared from the arm of a seven-month-old baby, because in such a young child explanations such as 'auto-suggestion' and other psychosomatic factors would not normally be considered relevant. The medical consultant involved with the case sent me copies of his reports on the baby, which showed that the tumour had measured 2.7cm in length and 16.1cm maximum circumference on 11th September 1987 but was "not clinically detectable" (i.e. had disappeared) by 5th October — just after a Christian woman had spent a whole day praying for the baby.[10]

The consultant was unable to offer any medical explanation for the tumour's disappearance apart from calling it "spontaneous remission": this is a rather loose 'catch-all' term which essentially means that it disappeared for reasons unknown to the medical profession. However, David Wilson, Dean of Postgraduate Medical Education at the University of Leeds, investigated on my behalf the available medical literature on this type of tumour (infantile fibrosarcoma) and discovered that there were no other recorded cases of so-called spontaneous remission for this cancer. Documented cases in which a child had not died but had survived this form of cancer were all those in which either the tumour had been cut out or the limb had been amputated, sometimes followed by chemotherapy or radiotherapy.[11]

Other 'healings' in Russia

Owing to my earlier research on 'miraculous' healings, and the kinds of facts cited above, I am probably more inclined than some anthropologists to attach a degree of credence to at least some of the reports related to me. Of course, in this I am also influenced by my own religious faith, which, as argued in the preface of this book, can actually be an advantage in conducting research on such topics. Certainly the fact that I had written a book on the subject often meant that people were more open to telling me about their experiences of healing.

However, it also meant that I was sometimes considered to be an 'expert' and invited to pray for the sick myself. This was 'participant-observation' of a different kind from that of many anthropologists! Soon I learned through experience that it was far better for people to be trained to do it themselves than to think that they had to wait for a Westerner to come.

Often there has been little or no obvious improvement at the time, but sometimes I have been surprised at the reports of later improvements. One case which stands out in my mind concerns an eight-year-old boy named Leonid whose right foot was so deformed that he could hardly manage to walk a few steps by himself. I was asked by his grandmother to pray for the boy, but rather than do so alone I invited the grandmother and another lady to join me in praying. This was in September 1995. The following January I was again visiting Kazan and the grandmother invited me to accompany her to the 'sanitorium' where Leonid was staying. Just as we arrived she mentioned that the doctors had been amazed at the improvement in Leonid when he was next examined after that session of prayer. As soon as Leonid heard that I was there, he ran from another room to greet me and jumped up into my arms.

However, this was by no means a total healing, as he still had some residual problems. Nevertheless, in October 1996 the grandmother showed me X-ray photographs taken of Leonid's legs at various ages. She claimed that the earlier ones showed that the bones in his lower right leg were noticeably shorter than in his left leg but that the most recent X-ray showed that, subsequent to the session of prayer, the right lower leg had grown to the same size as the left leg.[12] For almost two years I had believed that this was a genuine 'miraculous' healing, even though it had been partial and Leonid was also receiving conventional medical assistance. However, in July 1998 I heard from other people that it had been necessary for Leonid to have an operation because one leg had grown longer than the other. Was this therefore a case of a 'temporary' healing (whereby the leg had at first grown but then had ceased to grow at the same rate as the other one) or had we misunderstood the medical facts? Questions of this kind beset many claims of 'miraculous' healings.

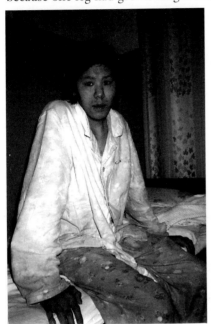

Tolya at the hospital in Anadyr

Another unanswered question for religious people is an inability to explain why God seems to heal some people but not others. One clue is given in the New Testament, when Jesus selected only one man for healing but apparently left others as they were: he later explained that he does "only what he sees the Father doing" (John 5:1-14, 19). Now some Christians in Russia also try to discern "what the Father is doing" — which might involve prayer for physical healing. I encountered an example of this is Chukotka, when some Ukrainian Pentecostals took me with them to visit patients in a local hospital. There I was introduced to a Chukchi man named Tolya, who said that he had been paralysed and unable to get up until one of the Christians prayed for him. Tolya felt the power of God come upon him and the paralysis disappeared. However, he was not at that time healed of his pneumonia and was therefore still hospitalised.

Although I do not dismiss claims of supposedly 'miraculous' healings, I also know that it is often difficult to obtain medical confirmations of them. A tendency towards exaggeration cannot be ruled out in some cases, so I tend to pay more attention to accounts given by people I know and respect, such as Julia Borisenkova, who with her husband Sergei founded the Cornerstone church in Kazan. She said:

"On 22nd February 1994 we were due to have our first service specially devoted to prayer for healing. Then we discovered that the public hall we were renting was due to be used by a psychic from 4pm until the church service at 8pm. Therefore some members of the church decided to attend the psychic's meetings in order to sit at the back and pray.

"When the psychic began to demonstrate his 'powers' and his methods of work, at first we began to pray against it. Then we felt that the Lord was telling us to praise and worship. From that point onwards the psychic's 'powers' began to fail. For instance, through attempted hypnosis he said to one subject, 'you will not be able to stand up', but she stood up immediately! Another man was told, 'you will not be able to separate your hands', but he did so straight away with no difficulty. The psychic was puzzled and tried to make excuses for his failures, saying that perhaps he or his disciples were tired. He said that these things usually work but he could not understand why they were not doing so on that occasion.

"After he had finished, one of our church members went out to the front and announced that there would now be a healing service. Many people who had come to see the psychic stayed on for the Christian service and several of them also reported healings. For instance, one woman had been unable to move her feet and had already had surgery for the problem but after receiving prayer she not only began to walk but even found she could dance. Another woman had a tumour in her womb but she felt it disappear when we prayed for her. Several people with high blood pressure found that it was reduced: one man's pressure had been 220 but after the service it fell to 120."[13]

It is easier to believe accounts by those whom I know personally and who are not prone to exaggeration. This is not possible in cases where the character of the informant is unknown. However, sometimes independent witnesses are able to confirm the salient facts. Therefore I sought such opinions in attempting to follow up the reports of healings claimed by evangelist David Hathaway. Anecdotal evidence indicated that some of the healings were sustained for at least some time after his meetings. For instance, Iida Lukina in Neryungri told me that a lady named Lyuda had been healed of "a problem with the woman's part of the body." Ten days after Hathaway's mission, Iida met Lyuda again, who confirmed that the healing had continued. Other cases from Neryungri reported by Iida included a formerly deaf girl who can now hear and also a woman who had previously been unable to walk unaided but can now walk by herself. It is not easy to obtain documented medical evidence for many claimed healings, but it is also difficult to dismiss such reports about disabilities for which normally one would not expect such dramatic improvements.

What is much more problematic, however, is the manner in which reports of such healings are used as a form of promotional material by Western evangelists, whose reports are often couched in a very dramatic style, focusing on the apparent 'success stories'. Such reports give the impression that everyone with a disability received healing. There is no mention of anyone who went away disappointed at not being healed, or who later suffered a recurrence of the disability. It would be wonderful if this were indeed the case.

Miracles

Usually, however, there are some people who receive healing but others who do not. For example, Pavel Bak, a Pentecostal pastor in Yekaterinburg, said: "I was injured during my military service and my legs paralysed, but, in answer to prayer, God healed me and enabled me to walk again. Once I prayed for a man with a similar condition, who was also paralysed, and God healed him too. On another occasion I prayed for a man with limited sight: God also healed him, so that he could throw away his glasses. But what I don't understand is why God seems to heal some people but not everybody."[14]

Pavel Bak in Yekaterinburg

The very nature of a 'miracle' is that it is an exception, but its occurrence does not guarantee that others will believe even if they see a 'miracle': for instance, when Tanya Osintseva was summoned to an 'interview' with the KGB, the officer commented to her that he might believe if he were to see a miracle. At that very moment they looked out the window and saw snow falling — from a clear blue sky! Nevertheless, the man still did not profess any belief in God.

However, the more dramatic cases of alleged miracles are the ones which stand out in people's memories. These 'exceptions' to the norm can take many forms, including the provision of material aid or rescue from a difficult situation, but these often depend on subjective assessments of unexpected timing or circumstances. Therefore in this chapter I have tended to focus on claimed miracles of healing because they are more tangible as case studies. However, many of the cases of healing which I have encountered involve relatively minor disorders — but ones which are nonetheless real — so that it is difficult to obtain any kind of medical verification. Were merely 'natural' healing processes at work when I prayed for a very deep wound in the hand of a Bashkort young man to heal up quickly? I thought it would need stitches but he had refused to go to a doctor because he assumed that they could do nothing to help. However, a few days after the accident he wrote to me that the wound had healed up far quicker than he would have expected.

A sceptic could easily dismiss cases like this as merely the subjective interpretation of natural processes. Even more difficult to assess are situations involving mental abilities. For instance, in January 1996 a Baptist family asked me to pray for their five-year-old son who was still unable to speak apart from a few words. Three months later I was told that the boy had begun to speak in sentences. Was this an answer to prayer or merely 'coincidence'?

Certainly physical healings do occur at times. In meetings like those led by David Hathaway, they are important as tangible proofs of the reality behind what is being said. What is also important, but is less dramatic and obvious, is the healing of repressed emotions such as anger, rejection and shame which can also express themselves in physical illness. The term 'psychosomatic' merely acknowledges that there is a close link between what one might call the 'soul' (in Greek, *psyche*) and the body (in Greek, *soma*), but it is often much harder to bring healing to the *psyche* than to the body.

For instance, in Yekaterinburg in 1990 a woman mentioned how she had been suffering from recurrent nightmares. They had started when Nadia had watched on television a programme by the psychic or occultist named Kashpirovsky. Nadia said she had been unable to draw herself away from looking into his eyes, even though she described them as "evil." As she watched the television, her head began to gyrate. Since that time she had suffered from nightmares. I prayed with her and invited God's Holy Spirit to come to her, to bring her peace. As far as I could see, nothing obvious was happening when, in the name and authority of Jesus Christ, I commanded any evil spirit connected with Kashpirovsky to depart from Nadia — but later Nadia told me how at that time she had felt something oppressive lift off her shoulders. The following year she told me how her sleep had improved from that evening onwards.

More recently, on a visit to Kalmykia, I was present at a meeting of Kalmyk Christians when one of them asked prayer for the healing of painful ovaries (initially described as "a woman's problem"). The leader of the meeting prayed along the lines of "God bless this lady", but when I was invited to pray for her too I felt that it was necessary to be more specific, as I suspected that it was due to her having had an abortion. It was not the kind of question to ask directly, so I asked if she had ever been pregnant. She then admitted to having had an abortion. This led into prayers of repentance regarding the abortion and her relationship with the child's father, before we committed the spirit of the aborted child into the hands of Jesus. Finally, using the name and authority of Jesus Christ, I commanded any spirit of death to leave her body. Immediately she felt a movement inside her, away from her womb up into her chest and finally out of her body.

Many other instances could be cited, but I described this one in a little more detail because it illustrates a need for healing of the whole person, not just the physical aspects. To do this takes time and a sensitivity to the person's total circumstances. At times very deep and painful memories might require healing — such as when a woman in Tatarstan needed counselling regarding her childhood experiences of sexual abuse from her father. In such situations it has been helpful to have the help of a local woman acting as a counsellor too.

Often I have found that my very role as an outsider has meant that people seem more willing to confide in me about personal issues than they would with their own friends or colleagues, or even with the pastor of a church. It seems to me that there is a serious need for counselling skills and a deeper understanding of the inner needs of people which are at present unmet by a majority of religious specialists, whether they be Orthodox, Muslim or Protestant. All too often what purports to be 'counselling' by religious people turns out to be little more than the quoting of scripture and telling people what they ought to be doing, with little feeling for the inner turmoil and hurt which also needs to be released.

So-called 'counselling' of this nature might be helpful for some people but it is directed largely at the mind rather than the heart. In Russia and Central Asia many people are carrying around inside themselves the 'baggage' of broken marriages, feelings of betrayal, distrust, anger, resentment and fear which can often dominate their lives. People are looking for love and acceptance — which are the real antidotes to feelings of rejection, distrust, anger and fear — but genuine love and acceptance can only be expressed through meaningful social relationships. In many cases religious organisations have absorbed some of the attitudes even of the Communist system in terms of addressing the mind rather than the heart. It is therefore not surprising if they also experience some 'leakage' through the loss of those who have not found real love and acceptance and continue their search for meaningful social relationships elsewhere.

Those who see religion in terms of formal rituals or the observance of regulations (such as times for fasting) are in danger of missing the real meaning behind those practices. Deeper than all of these outward expressions of religiosity are the real-life heartaches of everyday life and questions of meaning. Values and ethics are worked out in concrete situations of daily life where people know the ideals but need help to relate and apply them to the practical issues they are facing. Ordinary people often feel that religious institutions are irrelevant in terms of real advice or help in resolving the issues they face day after day. A truly caring, compassionate and loving social environment where people feel accepted and valued, and in which they are able to be real with one another, is far more important to them than a dead and formal religious tradition. To breathe fresh life into those dry bones will take a real miracle!

Perhaps it is for precisely this reason that many people feel attracted to smaller and more intimate forms of worship or religious teaching, such as the Orthodox house groups described in chapter seven. Unfortunately many of the more formal Orthodox services can often appear to be like staged performances as far as the casual visitor is concerned, so many parishes fail to incorporate people into a sense of community in which they feel free to discuss their personal lives. It is not surprising that people often feel more accepted and valued for themselves in some of the Protestant churches. Some Orthodox parishes or individual priests do also provide this same kind of pastoral care and interest in the daily lives of their parishioners, and these are the priests whose ministries seem most appreciated by their congregations. The next step would be the training and development of lay people who can also help to share this huge pastoral responsibility and thereby make the churches far more effective.

In this chapter I have focused on reported 'miracles', especially miracles of healing, but there are many other types of claimed miracles apart from those involving healings. Nevertheless, I have discussed healing miracles in detail because they are apparently more 'quantifiable' than those which involve subjective perceptions of apparent 'coincidences'. Certainly this same problem of interpretation can also apply to cases of healing miracles, but for medical cases there is at least some measure of the statistical probability of the expected outcomes and significant deviations from these might with some justification be referred to by the use of a term which Rex Gardner calls a 'convenient shorthand' — namely, the word 'miracle'.[15]

In practice, however, specific cases often involve several different dimensions of what might be labelled as 'supernatural' phenomena. For instance, people praying for healing often ask God for revelations or insights into the cause of a physical problem and seek divine guidance about the way in which to pray in that instance.[16] Often what are believed to be 'divine promptings' seem to take the form of a strong intuition — such as in praying for the Kalmyk woman with the painful ovaries whose previous abortion had not been divulged. A related category of revelation (sometimes referred to as 'miraculous' in religious circles) concerns the possibility of receiving revelations about future events. To some extent these are similar to the precognitive dreams already discussed in chapter two, but in my final chapter I wish to consider further the whole question of whether or not it is possible to obtain information concerning the future.

12
Facing the future

Premonitions

"When I was a child, in 1966, there were terrible floods in Ulaan Baator. At that time my father had taken my two sisters and myself to the celebration of *Naadam* [a traditional sports festival, especially involving horse racing and wrestling competitions]. We were watching the *Naadam* in the stadium but I began to cry and said I wanted to go home. My father wanted to watch the horses and my mother was away from home, doing concerts in the USSR, but I kept on wanting to go home. Finally we went home. Half an hour later there was a flood in which many people died, especially when a bridge near the river collapsed. My father said I saved our family."

This Mongolian lady said that her parents had not believed in God but that they began to believe in later life, after the age of 55, for reasons which she did not specify. I do not know whether or not it had any link with the fact that the informant's father had himself been sensitive to premonitions, as shown by another incident recounted by this lady:

"In 1967 or 1968 — I can't remember the exact date — there was a severe earthquake in Ulaan Baator. My father told me he couldn't sleep and he got up at 5am. He read the newspaper in the kitchen and smoked his pipe. Then he wanted to check me, and look in on me, but he not only went to my room but also took me outside the house. My father left me there while he then went back and fetched my two sisters and my mother. This happened in the winter when it was very cold. The earthquake happened five or ten minutes later, although in fact our house remained intact."

Behaviour of this kind demands a form of 'faith'. The premonition required action, even though there was apparently no rational reason for his actions. In some ways, one might say that such conduct requires greater 'faith' on the part of someone without a formal religion than for a religious person, because the latter could at least claim to be responding to the voice of God!

What is interesting is that many people claiming to be 'atheists', or to have little or no formal religious affiliations, also report premonitions and precognitive dreams which they interpreted as warnings about future events. I cited this particular Mongolian lady and her father in some detail because they were able to provide relatively specific accounts, whereas many others had forgotten the details of their premonitions or precognitive dreams. Both premonitions and precognitive dreams forewarn of future events, except that premonitions occur while one is awake and usually take the form of an intuitive feeling rather than a visual depiction of an event, as in dreams.

A 20-year-old Mongolian lady said that she has "quite a lot" of premonitions, such as: "Last weekend my boyfriend's brother was supposed to come to London from Ireland. We went to meet him but on the train I felt that he wouldn't be there, and he wasn't." (It turned out that he needed a visa

A street in Ulaan Baator, capital of Mongolia

from Ireland.) "Yesterday our lecturer in Economics didn't come. I was at home and had slept in late but the thought came to me that the lecturer wouldn't come. I got there ten minutes late and was told that the lecturer was sick so there would be no lecture."

Often premonitions take the form of a strong feeling or inner compulsion, as in the next examples: "When I was studying as a student [in Moscow], I remember one wintertime I was feeling very sad and upset without any reason. I was thinking about my grandmother [mother's mother]. At that time she was very old. I loved her very much because after I was born she used to look after me and my brothers. I couldn't bear to live so far away from her — she was very close to me, like a mother. I was very upset for a week, very worried, and decided to call Ulaan Baator to ask what was wrong. When I 'phoned, I discovered that she was ill. That was the beginning of the illness of which she died, over a year later. That was the first time in my life I lost a close relative who died: it's a very bad memory. I couldn't realise why I was upset [before I telephoned home], but I felt very worried that something was wrong with one of my relatives: I didn't know if it was my father or mother, or who. I still remember it, a very sad memory." (A man in his early thirties.)

"One day, ten years ago, when I was studying in Russia, I felt that my grandfather was dying. I had been very fond of him. When I phoned home and asked what was happening, they said, 'nothing; he's okay', but when I came home on vacation I discovered he really was dying." (A 37-year-old woman.)

A suburb of Ulaan Baator, consisting of yurts (called in Mongolian 'ger') surrounded by stockades

A similar feeling of unease was reported about a vehicle rather than a person: "Last year we had planned to go out to the countryside, to the opening ceremony of a big monastery named Amarbayasgalav. We went with British journalists from *New Scientist* magazine. However, I had a feeling about our vehicle, and asked the driver if everything was OK with the car. He said it was a good car, belonging to the Foreign Ministry — a four-wheel drive Peugeot, but 20km from the city, at the post for registering one's journey, he touched one of the wheels and it was very hot. Therefore we postponed our journey and left later that day after he went to have the wheels balanced."

When asked how he explained his feelings, this 42-year-old man replied: "I don't know where my feeling came from — probably a mistrust of the car, because I'd never driven in such a car before and I prefer Russian jeeps because we know they are very good in our countryside."

Some premonitions are closely associated with precognitive dreams or visions. Many people in Leninakan and Spitak, the settlements in northern Armenia devastated by the 1988 earthquake, believed that they had been warned about the disaster through visions of angels or precognitive dreams.[1] I have not interviewed any such people personally, but another example of a premonition associated with a dream or vision was recounted to me by a 34-year-old Kalmyk man, who said: "Three years ago I was seriously ill in hospital and the doctors were preparing me for an operation, cutting part of my lung because I had tuberculosis. While I was on the operating table they injected me with the anaesthetic but as I was falling asleep I saw in front of my eyes an image of my son. A great feeling of fear overwhelmed me. I cried out, 'Stop!' — I no longer wanted to have the operation. They stopped and let me go . . . Two or three days later I did have the operation and the lung was cut out, but the doctor who performed it told me that I was right to leave the operation that first day because the surgeon who was going to carry out the operation was not qualified enough to do it."

Part of a Kalmyk marriage ritual: the bride's mother haggles with the groom's relatives over the price for the bride's pillow and other bedding

When asked to explain how the image of his son produced this reaction, he said: "Seeing my son made me feel that way: I don't connect it; I just had the feeling that my son might be left alone without me. Before that time I used to drink too much, but after that experience I realised how much my son needs me so for almost three years I quit drinking and didn't drink at all." When asked how he interpreted his experience, he replied: "I don't know. It changes nothing if it comes from God or somewhere else: I'm absolutely not religious."

Premonitions are not uncommon, but often those who say they have experienced them have actually forgotten the details by the time they are interviewed. One might ask whether perhaps a sensitivity to premonitions might have been dulled among those brought up in a society which overtly pays little attention to these inner promptings. Might it be that children have a greater sensitivity to such promptings? This was the attitude of one nominally Muslim Tatar woman named Zillah, who related the following accounts:

"When Timur was three years old, in 1993, he kept saying we'd get a new flat near to a cafe called Shulpan where they sell ice cream. At that time we'd been living in a completely different district and didn't know that we'd get an apartment in the area near to that cafe, but in fact, at the end of 1993 we did move to that house.

"Then, in 1994, when we were coming back from Kazakstan, we stopped off in [the town of] Buguruslan to see some friends. Timur was then about four years old, and on the way he said, 'When we visit them, they'll certainly give us *pelmeny* [meat-filled dumplings] to eat.' Then, when we arrived at their home the housewife invited us to eat with them and in fact did give us *pelmeny*."

This second incident might be seen as less surprising, in so far as *pelmeny* is a relatively easy dish to serve up to guests who turn up unexpectedly, but the first incident is harder to explain away in 'natural' terms. Timur's mother related these accounts to me in 1997 but she was unable to remember more recent incidents of this nature: these two had stood out most clearly in her mind. On asking her opinion about how to interpret these events, she replied: "I think that children have a connection with the cosmos. [*Later she added:* "and with God"]. They are not yet sinners and a voice speaks to them from inside and they speak it out."

Responding to my prompt question about how adults lose this capability, Zillah replied: "If adults could do away with sin, they too would have the possibility of hearing God's voice also."[2] Zillah's sister Lilya agreed with her that children are less "sinful" (*greshniye*) than adults and saw a link between this and a receptivity to information from what they at first described as "the cosmos" but then added "or God." Perhaps the mention of God was included for my benefit, because their first reaction was to describe the information as coming from "the cosmos."

A 44-year-old Mongolian man, who had been married for two years but then divorced, commented: "My daughter surprised me once. I don't normally drink much, but once I did get very drunk and the police put me in prison overnight. Then, five days later, I got a letter from my daughter, who lives in the Gobi, and she wrote in it, 'Daddy, don't drink'. Letters from the Gobi take five days to arrive, so she wrote the letter on the day I got drunk."[3]

'Future-consciousness'

Can we know the future? To this age-old question there is an interesting perspective provided by mathematics. According to Maxwell's equations, it is possible for energy to travel not only from past to present and then to the future — that is, in the same direction as time as we usually experience it — but also to travel from the future to the past.[4]

If this is so, is such energy outside of our normal sensory range, like infrared light or the high frequency sounds which bats can hear but not humans? Might it be that mankind does have a potential receptivity to such information but it is manifested to us in the form of premonitions or precognitive dreams? Otherwise, are these experiences merely statistical quirks, whereby people happen to remember more clearly those times when their prior feelings or dreams happened to reflect events about to happen, but those feelings occurred simply by 'coincidence'?

At a more general philosophical level, there have been two principal interpretations offered regarding the source of any possible information from the future. More recent ideas have sometimes preferred the vocabulary of 'energy' or 'force' such as that which in studies of extra-sensory perception has been labelled as 'psi'. This kind of vocabulary has been the one which was allowed during the period of atheistic Communism, when it was virtually a 'taboo' to make use of vocabulary which spoke in terms of revelations from spiritual beings.

Use of the alternative, more traditional, vocabulary also tends to force the user into a choice, because in most religious traditions those spiritual powers have been further sub-classified into 'positive' (benevolent) versus 'negative' (malevolent) agencies. In popular parlance, are the revelations from God or from demonic powers? Religious traditions have developed a variety of tests for determining the source of such revelations — principally relating to the content, the context and the confirmations of the revelation.[5]

Often those who experience a premonition or a precognitive dream are uncertain about the source of the information. Even relatively non-religious people, however, still tend to use expressions such as "somebody up there must be looking after me." At an emotional level they tend to feel that the source of the revelation has a personality or consciousness and is also concerned about the human being who experiences the premonition or precognitive dream. This kind of comment can even be expressed by those who deny any belief in God if asked about it in a questionnaire-type survey separate from the context in which they experience the premonition or precognitive dreams. When they discuss such experiences they are more likely to mention their feelings of "someone watching over" them but such feelings are divorced from their professed beliefs at a rational level when discussing abstract ideas and beliefs in other contexts.

Facing the future

Those who prefer to use an apparently more 'neutral' vocabulary to explain precognitive dreams or premonitions in terms of 'forces' and 'energies' also tend to see those 'energies' as devoid of moral character. If they are merely 'forces' lacking personality, they are in effect sub-human, whereas the traditional religious viewpoint sees divine powers as not only 'super-human' but as also possessing moral attributes. Therefore the prophetic voice in human history is that which has seen human events as having moral consequences. The prophets of the Old Testament, for instance, repeatedly emphasised that history has a meaning, and that God's actions in salvation (e.g. the Exodus from Egypt, or the coming of the Messiah) or in judgement (e.g. political, economic or natural disasters, and the experiences of deportation and exile) are consequences of moral failures and lack of obedience to God.

According to this prophetic perspective, the future is conditional upon present actions. Morality also has a political and social dimension, in so far as judgement was often a consequence of political oppression or corruption. In this context, it is not surprising that repressive political regimes have by their very nature spawned religious movements which have looked to a divine retribution upon the oppressors.

To a large extent religious movements of this kind have been inspired by a third level of 'future-consciousness' beyond those of personal premonitions or religious perspectives on current events. Indeed, the latter are themselves usually dependent on this third level, namely the kinds of visionary experiences which are believed to be special revelations from God about the future. Not uncommonly significant dreams or visions form the basis for predictions about the future, but sometimes there is an uncertainty about the timing of events forewarned in dreams such as the following, which was related to me in Armenia by the 'Reuben' mentioned in chapter two:

"Three days ago I dreamed that I was in this country and it was very bad: someone showed me what would happen. It was horrible — very bad. I was asking Jesus to take me away but he was just hugging me and smiling and he said, 'Don't be afraid.' Then the picture changed: there was sea, a house, woods and a tree: very nice, with fresh air. Then a voice told me in English, 'This world will soon be ruled over by idols.' I didn't know the word 'idol' but I saw the word and later looked in a dictionary to see what it means."[6]

Some people report dreams which they believe to be warnings about the future but the events have not yet occurred. However, prior experiences of precognitive dreams which have then taken place in real life give such people greater confidence that they have correctly recognised certain dreams as warnings from God. A good example of this comes from a 45-year-old Russian Pentecostal woman in Moscow who in 1998 recounted the following:

"Some years ago, before I was a Christian, I had a dream about picking flowers in front of a certain building which I had not seen before. Years later, when I started work at the institute I looked up and recognised that right in front of me was that building I had seen in my dream.

"On another occasion I dreamed I was arguing with someone from the Orthodox church who asked me to speak in tongues. In my dream I refused because I knew he would make fun of it and that it would be mockery. A little later I was at a school when a discussion arose and suddenly the class became polarised between myself and a boy who was strongly Orthodox. He then asked me to demonstrate how I speak in tongues. Just at that moment I remembered my dream and realised that it had warned me of this, so I replied as I had done in the dream.

"I have also had dreams in which I believe Jesus was showing me what would happen in the near future. One of them showed a curfew on the streets of Moscow because Yeltsin was no longer there and there were to be elections for a new president. The trees were bare but there was no snow on the streets, as in late autumn or early spring. In another dream I saw the persecution which would break out against believers and which would be so serious that we would have to rely on God for even the smallest details of our lives. I saw the police break up a meeting and start to take away the tables but my husband went after them saying that they had no right to take away what did not belong to them; then one of the police officers turned round and shot my husband in the stomach. I also saw myself being taken to prison."

Near-death experiences

In some instances in which a powerful revelation is received the physical side-effects have meant that the recipient often falls to the ground or appears to be in a trance-like state.[7] Sometimes those who have been clinically 'dead' or very close to death have reported receiving at that time glimpses into future events on this planet.[8] Such experiences are apparently not as common as 'premonitions' but a Tatar friend of mine told me of a 'near-death' experience of his after he had fallen several metres and was lying unconscious on the ground. He saw himself in a place which seemed to be above the clouds, facing someone who appeared to be sitting on a throne — which Gali believed to be the throne of God. Gali was then given a choice of which woman he would marry: he did not see the women themselves but heard their voices, each one appealing to him to choose that one. He said that he chose the one who shouted the loudest!

A Turkmen woman doctor reported : "Twice I have almost died. The first time was when I was 25 years old and having an operation. I was given a very small amount of anaesthetic but I stopped breathing. At first I wanted to cry out but the surgeons couldn't hear me — no sound came from my throat. It seemed to me that I was going through a tunnel. I was aware of the surgeons and could see them looking at me but I couldn't see my own body.

"The second time was when I felt ill and had had some acupuncture treatment but still felt ill. I'd had a few drinks and then took a tablet to relieve the pain. A few minutes later I collapsed. This time I was aware of voices but didn't see anybody. There was no more pain and it all felt very peaceful. However, I then remembered my children and realised that they would have nobody to look after them and I wanted to go back."

Gayane, the wife of Khachik Stamboltsyan, is now in her early fifties but clearly remembered an experience which had taken place when she was 27 years old: "I was lying in the hospital bed very ill and in the morning woke up and thought that I had been dreaming. When Khachik arrived I told him that I had seen such a strange dream. In what I thought was a dream I had been lying in the hospital bed with doctors all around me. One of them gave me an injection, another was trying to do something to get my heart to beat. They were doing all these things and I was just watching them. To me it was very strange, because I was looking at them all around the bed — and I was in one place, one Gayane, and there in the bed was another Gayane! I felt very sorry for that Gayane who was lying down and suffering with the doctors all around her and looking so ill.

"Before that I had thought it was as if I was going out of the hospital room but it was dark there in the hospital corridor and it was as if I momentarily went out but immediately returned to the bed in the room. But I felt that something was waiting for me in the doorway. It was a dark, dark, black, black person awaiting me — and someone who was not pleasant. He was waiting for me and was speaking with me: he was not speaking in words but I was mentally connected with him. He was saying, 'Quickly, you have to be with me, quickly, quickly!' But I felt sorry that Gayane was lying there, whom the doctors were trying to help and were doing something for her. I said, 'I don't want to come. I'm sorry for her — I'll look and see what's up with her.' And then I just went on my knees on the floor and it was as if I began to beg, 'I don't want to come; I don't want to; I am so very sorry for her.' All the time I could see everything in the room — from here and there, from each direction: I could see it all. And then suddenly I felt that this person went away. Then it became easier for me; he had gone. After that I woke up and it was morning, and I thought it had been a dream.

"For fifteen years I didn't know what it was. They never spoke to me about what had happened to me. But then I met this woman who had been present at the time and she told me that I had been very seriously ill. It was then that I realised it had been a near-death experience."

At the time Gayane had no religious faith of her own. However, a few days after this experience she felt that she had to ask people to forgive her for all the wrong things she had done to them, no matter how apparently small they might seem, and to put all her relationships totally right. Behaviour of this kind seems to indicate that she felt her encounter involved a moral dimension of some kind. Even though she would have described herself at that time as an atheist, her experience prompted behaviour which in religious terms would be described as a profound and meaningful repentance.

A further implication is that her experience was felt to have been an encounter with a conscious being rather than an impersonal force. Whereas some premonitions and other experiences are sometimes interpreted as being due to some kind of presumed 'natural' faculty such as ESP or psi, those who have had a near-death experience often find this kind of explanation inadequate. Firstly, they find themselves perceiving the situation from a different vantage point from that of their human bodies. The fact that they often see the situation from above, somewhere near the ceiling, is an

Khachik Stamboltsyan, his wife Gayane and son Vartan

argument against telepathy from any other human being. Secondly, they sometimes describe conscious encounters with other non-material beings. Gayane, for example, felt she was in communication with a figure whom she described as dark and unpleasant, and who in religious terms would be described as a demon. Moreover, Gayane's later behaviour which resembles repentance would seem to indicate that at some deep level she sensed that it was very important for her to clean up her life and that her experience was felt to have moral implications.

At that time Gayane did not belong to any religious group, but three years later she discovered Khachik reading the Bible. When he said to her "there is a God" she felt an affirmation inside her, an instinctive reaction that this was true. One could speculate to what extent her reaction was influenced by her near-death experience.

Gayane's younger sister, Gohar, had also reported a near-death experience. She said: "In December 1982 I went to have a bath but there was a leak in the gas heating system for heating the water so I was poisoned by the gas. When I felt very bad I called my mother but when my mother came I fell onto her so she put me on the floor. I saw myself lying on the floor and Gayane and my father were trying to do something with me. But I was trying to tell Gayane, 'Leave me; don't do anything: it's very nice for me — I mean, a pleasure — and I don't want you to disturb me.' I also saw my mother running from one room to another because she couldn't do anything so she was trying to call the doctor to come and help me. When I was trying to tell Gayane that I was OK, I was touching her but she wasn't reacting to my touch or my speech.

"When Gayane was trying to do something to help, she checked my pulse and I thought, 'Oh, it means that I am dead', because I saw her reaction when she did this. I could see her face, but I was seeing it from above, so from this I understood that I was dead. I was in different places of the room, I could see everybody in the flat — my mother, how she was worrying, and Gayane, and I could see what they were trying to do.

"Nevertheless, I liked my situation. There was no weight, no problems, nothing. I felt freedom. At the time I was feeling that it was something very pleasurable: I felt free, and no problems; I could move very easily, and there was nothing to worry for. I didn't want to come back but afterwards I did come back and regained consciousness. I remember there was a lot of hysteria and excitement when I did come back!"

Some years later Gayane and Gohar were reading a book about near-death experiences when they together had another spiritual experience which could be described more as a vision. Gohar described it as follows: "In the summer of 1983, when my sister and I were reading [Raymond] Moody's book, we went to sleep in our parents' bedroom. There were two beds in there and we were in bed and reading the book when suddenly we saw a light at the door. It was very yellow, very bright, very warm, but this light was very strange because there was nowhere from which the light could be coming in that place: it was a light from nowhere! And even though I am much younger than Gayane, all the time Gayane was saying to me, 'Oh, go and see what's there!' It was frightening because the rest of the apartment was pitch black but suddenly there was this light in the corridor. It was circular, a ball of light — very distinct, with borders — but with darkness elsewhere in the apartment. The sphere was about 20 centimetres in diameter and the light was very distinct: beyond the light there was no light but instead it was very dark. The light was up in the air, as if hanging in mid-air: it was very strange. There was nowhere it could be coming from as the doors were shut, and so on. We saw it at about 3am and nobody else was in the flat. I too was afraid but there was nothing else for it, so I went to investigate. I went and looked, and returned to Gayane, saying 'There's nobody there.' She was still afraid and told me to go and look again. Then suddenly we felt no longer afraid and we continued to read, and then we went to sleep."

When asked how they explained the light, they replied: "Moody's book describes other people who had also seen spheres of light like this. However, we had only just started to read Moody's book when we saw that light, and had not yet got to the part where such an experience is recounted. So when later we read that someone else had seen a sphere of light like the one we had seen we believed what we were reading. It was a kind of sign that such a thing was possible."

Spiritual experiences of this kind can involve choices which have moral consequences and affect future actions. Often those reporting 'near-death experiences' have found themselves faced with choices about whether to remain in the other world or to return to their bodies. They have sometimes found themselves assessing their own lives and asking what they had achieved during their earthly lives which was of true lasting value and motivated by a concern for others or for God rather than a seeking after their own desires and ambitions.[9]

Individual 'near-death experiences' vary considerably in their contents, involving visions of hell as well as of heaven, and with varying degrees of intensity and detail. An early study by Raymond Moody amalgamated together features of several different accounts to form a conglomerate 'prototype' which is not necessarily typical of all cases and can therefore be misleading if regarded as a standard.[10] In particular, it does not take into account those cases in which the person finds him or herself being dragged unwillingly into a frightening and 'hellish' environment. Moody's research was initially inspired by the account of a professor of his named George Ritchie (to whom Moody dedicates his book): moreover, Ritchie's account is one of the few which actually conforms in most detail to Moody's 'prototype' in which one at first finds oneself outside one's physical body, then encounters a so-called 'being of light', experiences a review of one's own life and is then shown a glimpse of another world before returning to one's body. In Ritchie's case, the 'being of light' (identified as Jesus Christ) first reviewed the quality of his life on earth and then showed him various scenes from hell and then a glimpse of a beautiful city of light. At that time Ritchie had not read the final chapters of the Bible which describe heaven in such terms, but it is interesting that a similar pattern, in terms of the sequence of types of visions, is also discernible in the book of Revelation.[11]

Interpretations of history

Not only the biblical book of Revelation but also other apocalyptic writings have provided plenty of fuel for 'religious' perspectives on political events in Russia. Especially during the Communist period, it was easy to imagine that the military might of modern Russia was already described in the Bible in terms of "Gog, prince of Rosh [i.e. Russia]" who with his allies would invade Israel from the north (Ezekiel 38:2, 3; cf. Revelation 20:8).

Other interpretations of the Bible in terms of Russian history are less obvious but have an appeal in terms of their curiosity value. In 1990, when

Facing the future

Gorbachev was still in power, a student in Sverdlovsk tried to convince me that the Bible was referring to Gorbachev as the one whose head "seemed to have had a fatal wound" which had been healed and who was the "eighth king" (i.e. leader) of the Soviet Union whose capital is sited on seven hills (Revelation 13:3; 17:9-11). A further reason why Russia has been a fertile field for such ideas has been the ideology of Moscow as the "third Rome" (after Rome itself and Constantinople) which has been perpetuated since the 15th century, when Ivan III adopted the title of 'Tsar' (that is, 'Caesar').

Prior to Gorbachev there were already movements towards reform and the opening up of the former USSR.[12] However, it could also be argued that another major stimulus was the Chernobyl disaster, on account of which the government had to admit to weaknesses or mistakes and allow inspection of the site by outsiders. Many observers liken the name of Chernobyl — meaning 'wormwood' in Ukrainian — with the account in the book of Revelation of "a great star, blazing like a torch" which "fell from the sky on a third of the rivers and on the springs of water — the name of the star is Wormwood. A third of the waters turned bitter, and many people died from the waters that had become bitter."[13]

In 1984 many Christians around the world undertook to pray in earnest over the next seven years for religious liberty to come to the Soviet Union and for Bibles to become freely available. A mission named 'Open Doors', which proposed this "seven years of prayer," compared the former USSR to the biblical city of Jericho, which was tightly closed against the people of God but the walls of which collapsed after the Israelites had marched around it daily over a period of seven days. In 1984 it would have been hard for anyone to foresee the momentous changes which in fact occurred in the USSR during those next seven years.

A Polish Christian told me of a vision he had seen while on a visit to Moscow in 1992 shortly before the attempted coup to overthrow Gorbachev. While looking at Lenin's tomb in Red Square he received a vision of a huge person standing on top of the tomb. This figure toppled forwards and almost fell off the tomb but then regained its balance. The Polish Christian interpreted the vision as depicting to him the spiritual power behind Communism which had been knocked down but had not been completely ousted — and was seeking to regain its position.

Communist ideology itself provided a view of history which claimed to be comprehensive. It painted a scenario in which man had come from a relatively idyllic state of so-called 'primitive Communism' through the various historical stages of slavery, feudalism and Capitalism towards Socialism. This schema was subject to a number of different interpretations, in particular regarding societies such as Mongolia which were said to have 'bypassed' Capitalism and progressed directly to Socialism.[14]

This view of history as having a direction and purpose within a fixed succession of stages is actually very close to a religious conception of history.

A simplified rendering of aspects of both Marxism and Christianity could say that both see history as having started with a relatively idyllic simple form of society which had later become corrupted through human selfishness of one kind or another but was heading towards a future 'utopian' condition. There are practical consequences because here and now we have the responsibility to facilitate the progression towards this idealised future state.

However, predictions of the future tend to take one of two forms — either conditional prophecies or immutable ones. Often the predictions of disaster by the biblical prophets were conditional and could be averted if the people turned away from their corrupt practices or other forms of sin. On the other hand, there were some predictions of the future course of human history which were immutable. Two of these were given through dreams to monarchs: the seven years of plenty and of then of famine depicted by the Egyptian Pharaoh's dream, and the future course of history shown to Nebuchadnezzar. In both these instances, the interpretation came through another person who had a close relationship with God (Genesis 41; Daniel 2). It appears from the material presented in chapter two of this book that many dreams today also seem to reflect a consciousness of the future but the meaning of the dreams is not always apparent to the dreamers.

A Russian Orthodox monk is reported to have received a prophetic revelation in 1911 outlining the course of history in the 20th century (and beyond). If genuine, it appears to be an 'immutable' rather than 'conditional' form of prophecy, although the reference to Britain being "saved by praying women" has been to some extent self-fulfilling by prompting in Britain a prayer movement among women. However, the main points of the Russian monk's prophecy are worth quoting more fully because of its perspective on 20th century events in Russia and elsewhere. He said: "An evil will shortly take Russia and wherever this evil comes, rivers of blood will flow . . . It is not the Russian soul, but an imposition on the Russian soul. It is not an ideology, or a philosophy, but a spirit from hell. In the last days Germany will be divided in two. France will just be nothing. Italy will be judged by natural disasters. Britain will lose her empire and all her colonies and will come to almost total ruin but will be saved by praying women. America will feed the world, but will finally collapse. Russia and China will destroy each other. Finally Russia will be free and from her believers will go forth and turn many from the nations to God." [15]

The Armenian earthquake: questions of interpretation

An unsettling aspect of prophecies of this kind (and of some other 'religious' interpretations of history) is the attempt to say that specific events here and now reveal God's supernatural intervention in human history. Behind the surface appearance of events, the believer has to ask: "What is God doing?" What the Old Testament prophets saw as a form of divine judgement —

Facing the future

Ruins of some of the buildings — including a residential building (top) and church tower (above) — which remained standing after the December 1988 earthquake in Gyumri (Leninakan)

namely, the deportation to Babylon — in retrospect became a means to purge idolatry from ancient Jewish society. From a similar perspective, seventy years of Communism in the former Soviet Union might also be seen as a process of breaking down corrupt forms of religiosity and helping to make people see what is of more lasting value.

As a response to suffering, however, some people become angry or bitter, whereas others become more compassionate towards others going through similar difficulties. One particularly delicate area concerns the interpretation of natural disasters as signs of God's judgement. It is difficult to make such claims without appearing to be insensitive to human suffering. In 1988 this became an extremely sensitive issue in Armenia after the earthquake which devastated the town of Spitak and destroyed large parts of Gyumri (at that time called Leninakan). Many died, many others suffered; there were a few who reported 'miraculous' deliverances.

For instance, a woman who had been on the seventh floor of a building when it collapsed described how she landed on the ground unhurt, and said that she felt "as if something or someone carried me down." An old man on the ninth floor of a tall building sought God in prayer at the time of the earthquake; all the surrounding buildings collapsed, but his did not.

Other Christians, however, were trapped in the fallen buildings, but still had some stories of what in retrospect were interpreted as divine help. A woman named Rita tried to come out of her flat by a staircase while the building was collapsing around her. A large panel fell down on her, and she was trapped under it. Unable to move, she lay in a "crucified" position, her arms splayed out on both sides of her. When she heard the voice of one of her neighbours, who was also trapped, she told this young man to pray to God. As far as she knows, he did not do so, and he died. Rita, however,

The single room of a hostel inhabited by a family of eight whose home had been destroyed in the Armenian earthquake of 1988

was constantly in prayer except for the times when she lost consciousness. For several days she lay alone under the heavy panel, unable to move. She became aware that one of her hands had been wounded but she could do nothing about it. Then she saw an animal on the panel. Thinking it was a cat, she told it to go away, but then she realised it was a rat. She felt the rat near to her wounded hand, but then she lost consciousness again.

When the rescuers eventually discovered Rita, it was clear that the rats had eaten some of Rita's flesh. She was taken to a doctor who wanted to amputate

Rita's injured arm because the blood supply to her hand had been cut off during those days when she had been unable to move it and the blood in her hand had turned black. This thrombosis would have become infected, leading to gangrene.

Rita's father-in-law, however, was present and said to the doctor: "Don't do it: we don't need her without a hand. It's better for her to die than to live without a hand." The doctor acquiesced, and did not perform the amputation. However, Rita's hand did not develop gangrene. During the following weeks she realised that when the rats nibbled at her flesh they released the coagulated black blood from her body. On closer examination of her hand, it appeared that the rats had not punctured her vein, but only took out the black blood which would otherwise have become gangrenous. She believes that God sent the rats to preserve her hand. At first Rita had some residual immobility in that hand, but now it is completely restored and healthy, so she is able to use it normally.

Many of those who were asking the question "If there is a God, why has he allowed this disaster to occur?" were reminded of what had happened to the city of Ani, which at one time was the capital of ancient Armenia. Located about 30km south-west of Gyumri, within what are now the boundaries of Turkey, the city had been very rich but had not heeded the Catholicos (the Primate of the Apostolic church) when he called them to repentance. There was then a plague and an earthquake, so the city was later abandoned. Similarly, in Soviet times Leninakan had been regarded as a prosperous but materialistic city where the people were proud and felt no need for God; the Brotherhood movement had been unable to establish a branch there. Nevertheless, there were accounts of people who prior to the earthquake had experienced significant dreams or had seen visions which they believed were warnings of the coming disaster. After the earthquake many turned to God and became Christians because the disaster had shaken not only their homes but also their values and basic assumptions about the meaning of life.

A senior Armenian cleric (who preferred not to be named) told me that he had been accused of being "unpatriotic" when he preached that the earthquake was a judgement from God. Perhaps his experience is that of a prophet who is without honour among his own people. Nevertheless, his was not a lone voice because both within and outside of Armenia similar conclusions were being reached by a number of others. On hearing news of the earthquake, I too had asked God why it had occurred: as if in answer to my question, immediately into my mind came some passages from the biblical prophecy of Amos, who prophesied two years before a major earthquake in Israel. Amos commenced his message by speaking of the atrocities committed by Israel's neighbours — acts which in today's terms might be described as genocide or massacres. (One of the atrocities described by Amos was the ripping open of pregnant women by Israel's eastern neighbours: such acts were paralleled in 1988 during the massacre of Armenians in Sumgait,

Azerbaijan.)[16] Amos said that God's judgement would come on such nations too. But the focus of his prophecy was on the people who were supposed to be "God's people" — Israel — whom God was going to judge on account of the corruption, injustice and exploitation in their own society.

Corruption was in fact responsible for most of the death toll after the earthquake in Armenia, rather than the earthquake itself. The Armenians had good safety regulations but these were circumvented by bribery and corruption so that the buildings became death traps: for example, cement was siphoned off for personal use and a higher proportion of sand was used in making the concrete than should have been. Despite the region being known to be subject to earthquakes, tall buildings were constructed in violation of official regulations. There would have been far less loss of life if an earthquake of this magnitude had occurred in Tokyo or Los Angeles — and in Armenia the buildings which best survived the earthquake tended to be mainly older ones built prior to the Khrushchev era, whereas many of the more modern ones collapsed.

On my visits to Armenia I have not only brought with me what I could in the way of humanitarian aid but have also had opportunities to speak about these issues at churches, in interviews with journalists and on television. While recognising the immense tragedy which has occurred and the need to care for the injured and bereaved, I have also encouraged people to see the earthquake as an opportunity to make a fresh start and to turn away from the corruption which had brought so much suffering to their society. An even more difficult message to bring has been that of being willing to forgive one's enemies, because resentments and anger have become like a cancer within the hearts of the Armenian people. When in 1998 I spoke about this with Kassia Abgaryan, press secretary for the president, Robert Kocharian, she remarked that she would not want to give these up because the sense of grievance has become so much a part of Armenian national identity.

However, anger at corruption and injustice exploded on 27th October 1999, when the prime minister and other members of the Armenian Parliament were killed in an attempted coup by people who said they could no longer tolerate corruption. Two months previously the tragic results of corruption had been seen also in Turkey, where thousands of people lost their lives in a major earthquake because of the circumvention of safety regulations by corrupt officials and businessmen prepared to risk human lives in order to make money.

Concluding comments

A perspective such as this one about the Armenian earthquake is almost bound to be controversial precisely because it impinges on politics. Religion is considered to be 'safe' as long as it is restricted to private beliefs and practices, but it is controversial when it questions political policies or practices. Political acts are nevertheless reflections of the values held by those holding power — and those values are potentially influenced by religious teachings. However, religious values such as truth, justice, mercy and love can challenge the political actions of states of a more 'autocratic' or 'despotic' nature which throughout history have therefore tried to control religion in one way or another — either by co-opting religious leaders into the service of the state, or else by legislation, manipulation or persecution.

The problem for such states is that there are many different dimensions to religiosity. Formal, organised religion is merely one of these dimensions but is much more noticeable and might seem to be easier to control. Usually legislation is focused on this single dimension of religiosity, largely because other dimensions are much less obvious. However, that does not necessarily mean that they are less powerful.

In the earlier chapters of this book I have looked primarily at aspects of spiritual experience. To some extent experiences such as significant dreams are not necessarily regarded as 'religious' by those who report the experiences, but such experiences do prompt people quite often to reflect on issues which might be seen as 'religious'. Indeed, dreams and visions have also played an important role in many religious traditions. The problem for legislators is that experiences such as significant dreams, premonitions and so on are often unsought and spontaneous. It is impossible to legislate against such experiences. Moreover, experiences of this nature are reported not merely by those belonging to religious organisations but also by ordinary representatives of the general public. Atheists, agnostics, Muslims, Buddhists and others all report such experiences. In fact, so common are they in the general population that it is more likely that a person has had such an experience than otherwise. Studies in Britain, the USA and elsewhere have shown that between a third and two-thirds of the population report such experiences. Those who deny having had any spiritual experiences at all are often younger people, whereas those who have lived a little longer are more likely to report having had a spiritual experience at some time during their lives.[17]

Perhaps this was also a contributing factor in the demise of atheistic Communism. It is true that often people kept quiet about such experiences for fear of being laughed at or thought to be mentally unbalanced, but it is difficult — in fact, probably impossible — to repress the spiritual side of mankind for too long. Throughout history and all the cultures of the world there have been practices and beliefs which can be regarded as constituting some kind of religion, so that it appears more 'natural' to believe than not to believe. However, the dimension of spiritual experience is only the beginning. People then begin to ask questions about the meaning of that experience: "Is there a God, or some other spiritual being, who is communicating with me?" "If so, what practical consequences are entailed?" "Do I have the courage to do what this Being is asking of me?"

One source of religious belief is an attempt to make sense of personal religious experiences. Other sources are those derived from the revelations given to others (e.g. prophets or religious teachers), but these themselves are often based on spiritual experiences of some kind, such as visions and significant dreams. Such revelations are often written down and included within the canon of scriptures which form a further source of religious belief. Scriptures may also include revelations of God's activity or person as perceived through historical events or through the incarnation of God as a human being.

A third dimension to religion is practice. Religious acts are often expressions of beliefs, but in some instances beliefs can also be rationalisations of existing practices. However, it is likely that very often the beliefs also arise directly or indirectly out of personal experiences, and that associated practices are also responses to those experiences. Such practices can include not only obvious religious rituals but also ethical conduct based on religiously motivated moral values. Religious practices very often take the form of group activities. This is only natural, because mankind is inherently social. The very fact that mankind has managed to adapt to life in many different environments — including the hot deserts of Turkmenistan, the steppes of Kazakstan and the Arctic wastes of Chukotka — has been largely through the fact that much of our behaviour is learned from those around us rather than being determined by genetics. This gives a special richness to cultural diversity; it also means that it is important to respect and to learn from those like the Chukchi and Khanty who have become accustomed to living in harmony with nature even in relatively severe environments. Their corporate or individual religious practices and outlook can be shaped by their environments — as described in chapter three in terms of particular dangers in those regions — but the general motivations such as fear, memorialism and so on are found in many different cultures. It is the particular expressions of these motivations which are more specific to certain cultures, which in turn are influenced by the demands of survival in particular environments.

A common Russian attitude towards new religious movements which have emerged in recent years is that they "spring up like mushrooms." Just as some mushrooms are poisonous and dangerous, so it is necessary also to be careful about some religious sects and cults. However, there is a danger of thinking that one can control their growth merely by "picking the mushrooms" once they have grown, without recognising that they spring from complex and deep roots and that their spores can easily spread to other areas. To a large extent one might compare those spores to spiritual experiences which seem to arise from unexpected sources — and to grow rapidly in new environments.

Russia is far from being an 'Orthodox' or even a 'Christian' country. In Moscow it is estimated that about 150,000 to 200,000 people — that is, no more than two per cent of the population — visit the city's Orthodox churches for major festivals such as Easter. Moreover, many of those who consider themselves to be 'Orthodox' profess beliefs which are far from orthodox: according to a survey conducted in 1996, only 65% of those calling themselves 'believers' expressed a belief in a life after death and only 44% believed in the resurrection of the dead (which is a doctrine contained in the Apostles' Creed). At the same time, 29% of them believed in astrology and 41% in reincarnation; in other words, they held unorthodox beliefs which are normally considered incompatible with being Christians.[18] On the basis not only of such surveys but also of the kinds of material presented already in this book, it can be concluded that the 'real' religion of Russia is not Orthodoxy, nor even Christianity; neither is it paganism, shamanism or any other recognised ideology such as atheism. Instead, the religion of most inhabitants of Russia is a folk religion which incorporates into itself many disparate beliefs from a variety of different origins. Some of these many beliefs and practices, for example, include: fear of the evil eye and of other curses, astrology and other forms of divination, rituals related to childbirth, memorial rites for deceased ancestors, fear of ghosts and the consultation of psychics, mediums and faith-healers. However, on top of the folk religion might be a thin veneer of practices associated with Orthodox Christianity which makes people think of themselves as being Russian Orthodox even if in practice their everyday lives have little or nothing to do with this religion. In practice they could be described as materialists with many folk beliefs.

The same could be said about the nominally 'Muslim' peoples of Central Asia and the Caucasus, except that the 'thin veneer' has an Islamic colouring rather than the Orthodox one found in Russia. Throughout the whole region there are local variations in the underlying folk religion, which over the centuries has absorbed elements which originally might have been derived from shamanism, Zoroastrianism, Judaism, Christianity, Islam or various 'New Age' cults and beliefs. Even beliefs which in theory might appear to be incompatible with one another can be absorbed alongside each other into this underlying folk religion (or religions). The same applies also to Buddhism among the Mongolian peoples. Therefore it is a serious fallacy to label Russia as an 'Orthodox' country, or Mongolia as a 'Buddhist' country or one of the Central Asian republics as a 'Muslim' country. Any attempt to impose such labels on the country through legislation or coercion would be to repeat the mistakes of the Communists who tried to impose atheism by force. By legislation one cannot control the inner depths of the human spirit.

Throughout Russia, Central Asia and the Caucasus there are many people who yearn for a society based on the values of truth and love — a society in which honesty becomes the norm in commerce, justice is maintained in the courts, integrity in politics and faithfulness in marriage. In practice, however, they feel pressurised by circumstances into doing those things which ideally they would prefer not to do. Those who do try to live more honestly, even at a sacrifice to themselves, are often those who are motivated by religious ideals. In some cases their willingness to speak out against corruption and injustice means that they have been subjected to harassment and persecution by those in positions of authority whose power is maintained by corruption or lies.

However, any economy or political system which becomes permeated by bribery, extortion, injustice and violence will find that its own foundations become undermined by these same factors: not only individuals but also organisations and governments will reap what they have sown. Just as corruption was responsible for much of the devastation caused by the Armenian earthquake, now in many nations of the region can be seen more overtly the tragic consequences of crime, drug abuse, the collapse of health and welfare systems and the impoverishment of large sections of the population. A programme of economic or social reform in itself is insufficient without a spiritual regeneration of the nation because the establishment of societies based on truth and on a genuine concern for others demands a return to values often associated with religions — such as honesty, social justice, compassion and a sacrificial giving of oneself for others.

At the dawn of the 21st century it appears as if religion is likely to become an increasingly 'hot' issue in the Caucasus and Central Asia. The persecution of religious groups in Turkmenistan, Uzbekistan and Chechnya, with harassment also in Azerbaijan and elsewhere in the region, has been frequently reported during 1999 by the Keston News Service (produced by the Keston Institute in Oxford), and human rights monitoring groups such as Human Rights Watch and Amnesty International, as well as the US government's reports on human rights in countries of these regions. Such persecution includes the imprisonment of alleged Islamic fundamentalists (so-called Wahhabis) in Uzbekistan, the destruction of a temple belonging to Hare Krishna devotees in a village near Mary in Turkmenistan, the confiscation of religious literature, trumped-up charges against active Christians and the beheading of a Baptist pastor in Chechnya — with his head put on public display in Grozny as a warning to others. In countries like Uzbekistan and Turkmenistan where no political opposition parties are allowed to operate freely, there is a reversion to the conditions of repression which had prevailed under Communism. However, persecution is almost always motivated by fear: autocratic governments fear the power of religion as a force which motivates people to seek for a new society based on principles of justice and truth.

It is also fear (of their own loss of power) which motivates some governments to ban or repress any political opposition, but the very process of repression can in itself polarise the situation and generate that which the government most fears. After gaining a monopoly of institutional power a single-party state needs to find a scapegoat to blame for its own shortcomings. The 'enemy' might be real or imagined, and could be either external or internal — or both. Ethnic or religious minorities have sometimes become such scapegoats, resulting in their being persecuted. A legacy of the Soviet repression of religious minorities remains today in the form of a minefield of burdensome regulations on the registration of religious groups in countries such as Uzbekistan. It also allows plenty of scope for abuse — and is a particular problem for those who, on account of their religious values, refuse to pay bribes to corrupt officials.

Facing the future

Whereas the Communists regarded an alternative economic system (i.e. capitalism) as a threat, what continues to be even more dangerous to a repressive regime or to corrupt officials is an alternative value system. What in religious terms might be called the values of the Kingdom of God in itself prevents some religious groups — for example, the Seventh Day Adventists whose church was destroyed by the Turkmen government in November 1999 — from responding in a violent manner to persecution. In other cases, however, the very process of marginalisation might push some people into responding with violence, as illustrated in 1999 by the actions of Muslims who had fled from Uzbekistan to Tajikistan but then took hostages in southern Kyrgyzstan in an attempt to put pressure on the government of Uzbekistan, where they wanted an Islamic state. Whether or not the perceived Islamic 'threat' to the government of Uzbekistan had previously been real or imagined, the very process of repression had fuelled feelings of frustration which exploded into violence. However, violence by agents of the government is not so obvious because it is not so easy to prove cases of alleged torture by police officials or the planting of drugs on people as an excuse for imprisoning them.

Persecution can often intensify people's desire for change. Either they will seek revolution, which in this case is more likely to be fuelled by fundamentalist Islamic groups, or else the government can choose to defuse this potential time-bomb through internal reform. An awareness that change is inevitable means that governments have a choice: either to seek their own short-term profit (not only by trading in merchandise but also at the expense of the lives and souls of their people) or else to seek the longer-term good of their nation by being themselves the instruments of reformation.

At the beginning of the 21st century, the social, political and economic environment of Russia is one which is tending to cultivate a greater sense of independence and self-determination rather than centralised control. It is not surprising if people should look towards expressions of their religious inclinations which allow self-expression and greater degrees of individualism. I would suggest that the Russian Orthodox Church in the past was very suited to a society which placed a high value on community and group action. However, in modern cities there is an increasing sense of 'anomie': basically, people feel isolated and, quite often, lonely, but are looking for social equivalents of the village. Sometimes religious groups provide that sense of identity and cohesion, but it tends to be found more often in the 'new' religious movements than in traditional Orthodoxy. Any religion which wants to survive as a relevant, vibrant force in the 21st century, rather than becoming a museum piece, has to be able to move with the times. This does not mean changing its basic beliefs and values but merely recognising that the outer expressions of those beliefs — the forms of worship and so on — need to be made relevant to a generation brought up in a 'new' form of culture, after atheism.

Handkerchiefs attached to a bush near a church in Armenia.

Notes

Introduction: Spiritual awareness

1. Following Forsyth (1992, p xx) I am omitting the apostrophe often used to designate the Russian 'soft sign' in names like Kazan (= Kazan') or Almetevsk (= Almet'evsk) because "its occurrence is obvious to those who speak Russian and it adds little for those who do not". Similarly, the English 'j' has been used instead of 'dzh' in names such as Tajik or Azerbaijan. Indigenous names of ethnic groups are now being preferred to the Russian terms, so I am generally using terms such as Kazak (rather than the Russian spelling of Kazakh), Bashkort instead of Bashkir, and Sakha instead of Yakut. I also use refer to the people or language of Azerbaijan as 'Azerbaijani' because Akiner (1986, p 105) describes the term 'Azeri' as "archaic": nevertheless, it seems to me that the term 'Azeri' has been enjoying a renaissance and nowadays is commonly used not only in French publications (where it had continued to be used) but also in English. Both terms are used in English on internet sites about the people of Azerbaijan: e.g. 'Azerbaijani' at http:/www.president.az/azerbaijan/ az7.htm but 'Azeri' at http://www.president.az/azerbaijan/ az8.htm. Citations of works in Russian or Udmurt are presented in English transliteration.
2. Lewis, 1998a, pp 70-71.
3. At the time this was named the Alister Hardy Research Centre but now it is renamed the Religious Experience Research Centre at Westminster College, Oxford.
4. Hay, 1982, 1987, pp 120-34.
5. Lewis, 1987a; 1987b; 1990, pp 112-20.
6. Ostrander & Schroeder, 1970.
7. Tylor, 1873, p424.
8. Southwold, 1978.
9. I am grateful to Sergei Kudrin for this observation about the Communists' use of the word 'glory' (*slava*): Sergei also compared this with biblical instances of rulers whose downfall had come after they had accepted glory and divinity for themselves rather than acknowledging God as the only one worthy of glory (Daniel 4:29-35; Acts 12:21-23).
10. Others, such as Binns (1979, 1980) have made the same observations.
11. For Japan, see Lewis, 1993a, pp 268-70. For Britain, see, for example, Towler, 1974.
12. The suppression of religion in the former USSR meant that opportunities for such counselling were denied to most people; nowadays relatively few members of the general population think of asking a priest or pastor for advice but some might instead consult a medium or psychic — and pay money to do so (cf. Boyes, 1998, p17).
13. For example, a discussion of Islam in the Soviet Union by Karklins (1986, p185-202) focuses on formal outward rites such as fasting and prayers, as well as rituals connected with circumcision, weddings or burials. These are also the kinds of rituals which could be studied by Rimsky-Korsakoff Dyer (1991) and Chylinski (1991). A very indirect indication of Muslim religiosity comes from statistics on the numbers of mosques or mullahs, but during the Communist period these had been severely curtailed and controlled by the authorities and the number of 'unofficial' Muslim clerics was hard to ascertain (Bennigsen & Lemercier-Quelquejay, 1979, p151; Atkin, 1992, pp 63-64). Other studies tended to focus on religious references in literature (e.g. Jeziorska, 1991), including that of previous centuries (e.g. Feldman, 1992). Often scholars have had to concentrate on material from pre-Soviet times, both for Islamic cultures (e.g. Algar, 1992) and also for Siberian shamanism (e.g. Stepanova et al., 1964; Vdovin, 1978).
14. For further details, see Lewis, 1987a; Lewis, 1987b and Lewis, 1990, pp 112-20.
15. Lewis, 1998a, p74.
16. For example, accounts of so-called 'Arctic hysteria' were full of descriptions of religious experience of one kind or another (Czaplicka 1914, pp 309-21; also in Czaplicka, 1998).
17. For details, see Lewis, 1989.
18. Schmidt, 1931, p6, cited by Evans-Pritchard, 1965, p121, and Southwold, 1983, p61.
19. Lewis, 1993a, pp vii-xi. However, more recent experiences have made me aware that there is an ethical problem involved when the researcher feels obliged to participate in practices which conflict with his or her own conscience. I am confident that all researchers — whether they have any personal religious convictions or not — do have a conscience of some kind and are not unlikely to face ethical issues of one kind or another in

the course of their research. Not so much the existence but more the content of any questions of conscience is likely to be influenced by whether or not one has a personal religious faith: for instance, researchers who are practising Jews, Muslims, Buddhists or Hindus would have questions of conscience if asked to participate in rites involving the eating of food prohibited by those religions. Similarly, anyone with religious convictions about certain rites being essentially demonic would struggle with their consciences if expected to participate in such activities: to some extent this happened during the Soviet period when some religious believers held such views about the atheistic Communist system itself and refused, on the grounds of conscience, to participate in organisations such as the Komsomol, with its associated rituals and obligations. On the surface one might expect that academic researchers would not be subject to pressures of this kind but sometimes the demands of etiquette and the expectations laid upon those conducting research through participant-observation can mean that researchers with religious convictions of their own do face questions of conscience. A Christian named Peter Horrobin expressed this to me in a rather blunt but clear way by asking: "If one is studying prostitution, does that mean one has to sleep with the prostitute too?" I am reasonably sure that most researchers, whether they have a religious faith or not, would concur that personal ethical standards can set justified limits on the degree to which the ideals of participant-observation can be implemented in practice.

20. The anthropologist Kenelm Burridge (1969, p4, footnote 2), for example, writes that not to 'believe in' phenomena such as trances, stigmata, possession, levitation, walking on hot coals without being burned or skewering the cheeks without leaving a wound is surely equivalent to being a 'flat-earther'.

21. See, for example, Platt 1976, pp 113-122, or Evans-Pritchard, 1951, pp 83-85.

22. Southwold, 1983, p8.

23. This is the 'uncertainty principle' named after the German physicist Werner Heisenberg (1901-76).

24. Southwold, 1983, p8.

1. Peoples and Cultures

1. In Chukotka and Kamchatka a prohibition on sales of alcohol on working days means that violent deaths by murder mainly occur at the weekends — most often being reported on a Sunday — whereas in Western Siberia such deaths are spread out more uniformly throughout the week (Pika, 1993, p69).

2. Pika & Bogoyavlensky (1995, p70) state that in the Yamal Peninsula 'the average life expectancy among the indigenous population is almost nine years shorter and the standardised index of mortality is 1.8 times higher than that of the nonindigenous population.' Lyudmila Ivanko, a sociologist who has also worked among the northern peoples, mentioned to me in 1991 that life expectancy among the northern peoples was about ten to fifteen years shorter than among the Russians.

3. Forsyth, 1992, p10.

4. A similar process occurred also among the Yukagirs, where Christian saints became functionally equivalent to the spirit protectors of birds, fish and animals (Vdovin 1978, p410, citing V.I. Jochelson Religiozniye verovaniya narodov SSSR [Religious Beliefs of the Peoples of the USSR], Moscow and Leningrad, 1931, p63).

5. Dewdney, 1993, pp 237-39.

6. Crisp (1991) provides detailed information on the degree to which Russian has tended to take the place of indigenous languages in Central Asia, Azerbaijan and Tataria, including also very helpful comparative material from Siberia.

7. Olcott, 1987, pp 103-4.

8. Sadikov, 1990.

9. Akiner, 1983, 1986, pp 285 & 326.

10. Forsyth, 1992, p27.

11. Olcott, 1987, pp 101-2; Akiner, 1983, 1986, pp 289, 301 & 337; Rohlich, 1986, pp 102-3.

12. Taheri, 1989, p116.

13. Russians (numbering 1,548,292) and other nationalities make up the remainder of the population. I am grateful to Ildus Ilishev of the Bashkortostan Supreme Soviet for his help in obtaining these population figures. Most of the other population figures cited in this book are taken from the 1989 census data as published by the Gosudarstvenniy Komitet SSSR po statistike [State Statistical Committee of the USSR] (1990) and the Gosudarstvenniy Komitet RSFSR po statistike [State Statistical Committee of the Russian Federation] (1990). In some cases the figures are also checked with those provided by articles on individual ethnic groups in V.A. Tishkov (ed.) *Encyclopedia of the Peoples of Russia* (1994), which often gives more detailed figures on the regional distribution of specific ethnic groups.

14. An indication of this is shown by a word search on the internet: in October 1998 the AltaVista search engine produced 37,040 matches for the word 'Kazakstan' (89,660 for 'Kazakhstan'), 118,042 for 'Azerbaijan', 40,970 for 'Turkmenistan', 39,240 for Kyrgyzstan and 33,870 for Tajikistan but only 3,677 for 'Tatarstan'.

Notes

15. Akiner, 1983, 1986, p57.
16. Quoted during an interview for a BBC Radio 4 programme broadcast at 7.20pm on 3rd December 1991.
17. Akiner, 1983, 1986, p71.
18. Akiner, 1983, 1986, p77.
19. Hostler, 1993, p37. He also notes that "nearly every remnant of the ancient faith has disappeared" among the Muslims, whereas the Russian Orthodox Chuvash "easily identifies Orthodox Christian saints with the old pagan gods".
20. I here use the term 'pagan' as a translation of the Russian word *'yazycheskiy'*, as an umbrella term for indigenous religions of Siberia and European Russia. I do not wish the term to be misunderstood as having any kind of perjorative connotations but the problem is that alternative terms such as 'primal religions', 'animism', 'shamanism' and so on all have their shortcomings and possible wrong connotations too.
21. Akiner, 1983, 1986, p102.
22. Gouchinova, 1997, pp 60-62.
23. Fisher, 1978, p169, 178-79; Sheehy & Nahaylo, 1980, p11; Allworth, 1988, pp 145-46.
24. Huttenbach, 1993, pp 66-67.
25. I am grateful to Rosa Shatayevna Jarylgasinova, an anthropologist at the Institute of Ethnography, Academy of Sciences, Moscow for this information on the Koreans of Central Asia.

2. Dreams & visions

1. For example, Genesis chapters 40 & 41; Judges 7:9-15; Daniel chapters 2 & 4; Matthew 2:12.
2. See Ostrander and Schroeder 1970.
3. According to the various contexts of the dreams, they can be interpreted as being from God or else from evil spirits.
4. Lewis, 1991a, p49.
5. This woman declined to specify the type of operation except that it was 'internal' — presumably gynaecological.
6. This man also commented that his wife practices 'chi-gun' (a Chinese martial art), and he therefore suggested that her frequent dreams or spiritual experiences were related to a spiritual power which she senses and which she describes as "a stream of energy."
7. Lewis, 1991a, pp 53-54.
8. Hay 1982, 1987, pp 125-26.
9. For further discussion of this topic, see the classic work by Mary Douglas (1966).
10. The same metaphor was used in an early Christian song quoted in the New Testament by St Paul. It contained the words: "Wake up, O sleeper, rise from the dead, and Christ will shine on you" (Ephesians 5:14).
11. 'NLO: poslantsy Satany' in *Golos Vsyelonnoy*, April 1991, pp 1-3.
12. Lewis, 1991a, p51.
13. This was his explanation of precognitive dreams but he had other explanations for different kinds of dreams. "How I sleep can also affect my dreams," he said: "so that if I sleep on a cold floor I dream of being in the tundra: that often happens. Also if I've been very busy before sleeping, my sleep is restless and I have dreams that someone is hitting or killing me."
14. Lewis, 1991a, p54.
15. Lewis, 1991a, p52.
16. Lewis, 1991a, p51.
17. Lewis, 1996b, pp 142-43, 147. For further discussion on the nature of these lights or 'balls of fire', see Lewis, 1993a, pp 66-70.

3. Fear

1. Lewis, 1996b, pp 141-42, 146.
2. I was not shown any of these, but in Russian the local people referred to them by the term *'idoly'* ('idols').
3. 'Feeding the fire' is commonly practised among many other Siberian peoples and the practice has sometimes been adopted also by Russian hunters in the north (Tugolukov, 1978, p425). In Mongolia a guest would honour the spirits of the felt tent (*ger*) by sprinkling mare's milk or putting the fat of a sheep's tail on the fire. Fire among the Nenets and Enets seems to have also represented bonds of kinship (Vasiljev, 1978, p435). A reverence for fire as a source of one's very existence in the Arctic is similar to veneration of the sun — a practice which was widespread not only in Siberia but also among peoples such as the Udmurts. An urban variant of such customs is illustrated by a Chukchi family in the town of Anadyr who put outside on

the windowsill of their upper-level apartment, as an offering to the spirits, a portion of the sweets I had given them as a visiting gift.

4. Lewis, 1996b, pp 140, 144.

5. Musk, 1989, pp 22-30.

6. For the Russians, the table was said to represent the palm of God's hand, and for the Udmurts God's throne: therefore both agreed that one should not knock or rap on the top of a table. (A previously widespread idea that a woman who sits at the corner of the table would not marry is now reversed in some circles to say that such a woman will get 'corners' — i.e. of an apartment, which is now hard to obtain privately: in other words, she will marry.).

7. Svetlana Nizovtseva, personal communication. She described this deliberate cursing in Russian as *"navesti porchu"* — the Russian word *porcha* (literally 'spoiling') rendering the Komi word *sheva*, a kind of curse.

8. Khan, 1998, p34.

9. Lewis, 1996b, pp 143, 147.

10. This same informant also mentioned that some Mongolians had reported seeing in cemeteries at night strange and inexplicable lights which he called a 'ghost light' or 'ghost fire'. I do not know how widespread such reports are in East Asia, but similar accounts have also been reported from Japan (Lewis, 1993a, pp 66-70).

11. I was told about this taboo by a well-educated Kazak woman in her thirties who had lived for many years in Russia but nevertheless refused to wear any kind of clothing previously used by another woman.

12. This is argued by Douglas (1966) as a generalisation applicable to many — perhaps even all — cultural contexts.

13. Hostler, 1993, p32-33; Rohlich, 1986, pp 37-47; Taheri, 1989, pp 66-67; Akiner, 1983, 1986, pp 57-58, 432.

14. Hostler, 1993, p33.

15. Hostler, 1993, p38.

16. For instance, the Muslims refer to Jesus as 'Isa' but the 'Christianised' Tatars use the Russian form 'Iisus'.

17. Taheri, 1989, p67, 79.

18. Musk, 1989, p29-30.

19. The same is almost certainly true of forms of Christianity elsewhere in the world, except that the types of cultural influences are different: in the West, for instance, there is much less emphasis on biblical teachings to do with honouring one's parents. Within Buddhism too, it appears that the emphases put upon certain teachings are influenced by local cultural values and concerns, as documented for Sri Lanka by Southwold (1983). In Japan it is widely recognised that Buddhism in practice is largely concerned with the ancestral cult — an East Asian cultural emphasis — rather than with the formal doctrines of Buddhism (Lewis, 1993a, pp 61-64, 179-96, 200-8, 216-21).

20. Likhachev, 1990, pp 434-35; Rowe, 1994, p7.

21. Proverbs 8:10 (cf. 1:7; Job 28:28).

22. Rowe, 1994, pp 3-4.

23. 1 John 4:18 (New International Version).

4. Death

1. Buckley, 1997, pp 47-51.

2. Vladimir Vladykin, personal communication; Filatov and Shchipkov (1997, p178) also mention high levels of suicide among the Udmurts.

3. Later the man explained that the spirit was thought to be that of the grandfather's brother, but in Russian he had initially used a term meaning 'grandfather' or, by extension 'old man'.

4. A probable contributing influence on these beliefs was the traditional Yukagir respect for ancestors as mediators between the human community and the realm of nature. Moreover, these ancestors were also regarded as having been shamans (Vdovin, 1978, p408). "The Yukagirs were unique among the peoples of Siberia in the degree to which the cult of the clan shaman was developed": after the clan shaman's death, his body was "cut into pieces, the flesh and bones being distributed among members of the clan as sacred relics, while the skull was placed on a wooden body as a spirit effigy" (Forsyth, 1992, p75; Vdovin, 1978, p409; Stepanova et al., 1964, p796). Though perhaps not as strongly developed as among the Yukagirs, there were nevertheless aspects of an ancestral cult among many other peoples of Siberia, including the Koryaks of Kamchatka and the Chukchi (Vdovin, 1978, pp 410-17).

5. Lewis, 1996b, pp 142, 146-47.

Notes

6. For example, Vasilevich (1968, p323) writes that "the shamanistic ability was regarded as heritable through the male line" but sometimes 'the spirits of shaman ancestors . . . could pass down through the mother's side . . . Succession might skip three or four generations and cover over 400km."

7. It was unclear whether she was referring to Russian atheists or Christians, but at least some Russian Orthodox Christians would agree with the teaching (now prevalent in many charismatic churches) that demonic spirits can be transmitted from one generation to another.

8. Tugolukov, 1978, p424.

9. Tatyana Vladykina, personal communication.

10. Filatov & Shchipkov, 1977, p180. They also report that the dowry consisted of a horse for a man and a cow for a woman.

11. Lewis, 1997a, pp 229-30.

12. Sometimes similarities are also explicitly pointed out, as illustrated by the comment of an Evenki man who told me that they visit their graves on the ninth and fortieth days and one year after a death, saying that this pattern is like that of the Sakha (Yakut).

13. In response to my enquiries, one man told me that there is no idea that the soul rides on the reindeer but another Yukagir man, aged 33, said that the spirit of the reindeer accompanies the dead person's spirit so that he or she can ride to heaven rather than going on foot. He also remarked that it is good if the sacrificed reindeer has convulsions because that means it is galloping to heaven. However, he also admitted that different people have varying opinions of the afterlife because they had all been brought up under Communism and there was no longer any common point of view.

14. Rogozin, 1993, p27. The only possible New Testament precedent for such a practice is the obscure comment in 1 Corinthians 15:29 in which St Paul refers to a practice of baptism on behalf of the dead.

15. There are suspicions that aspects of pagan imagery which had already been absorbed into Orthodoxy prior to the time of Peter I (the Great) have been preserved within the traditions of the 'Old Believers'. For example, the cross on the back of their priests' robes often has arms of equal length and is enclosed in a circle, which is reported to be a pagan form of cross and is certainly not the conventional Orthodox cross (Irina Trushkova, personal communication). However, some other possible parallels, such as the way in which the robes of the Old Believer priests are divided into three sections, representing three 'worlds' — this world in between the heavenly and the lower realms — are ambiguous because of similar concepts within both paganism and Christianity.

16. Musk, 1989, p135.

17. Akiner, 1996, p98. Chinese Muslims who have settled in the former Soviet republics also observe rites to honour or propitiate the ancestors. In view of the prevalence of ancestral rites in China, it has been argued that these rites are certainly pre-Islamic ones which the Muslims continue to observe (Dyer, 1991, pp 83-85, citing M. Shushanlo (1971), *Dungane (istoriko-ethnograficheskiy ocherk)* [The Dungans, a historical-ethnographic sketch] (Frunze: Izdatel'stvo 'Ilim'), pp 250-55.

18. Beliaev, 1990, p414.

19. Lewis, 1997b, p65.

20. Prokofeyeva et al., 1964, p536.

21. Alexander Chernyk, an anthropologist from Perm (personal communication), was present at a Komi-Permyak ritual which commenced with the announcement: "We are here to remember Adam and Eve and Boris Mokin." Chernyk did not know who Boris Mokin was; it seems to me that the reference to Adam and Eve was possibly intended to evoke a sense of continuity with the distant past and with one's ancestral roots.

22. The sixth of the Ten Commandments states the obligation to honour one's father and mother (Exodus 20:12) and this was further re-emphasised in the New Testament by Jesus (Mark 7:9-13) and by St Paul (Ephesians 6:1-3).

23. Nikitina, 1997, p12.

24. Nikitina, 1997, p51.

25. Pika, 1996, p55.

26. Chylinski (1991, p167) reports that circumcision in Central Asia has persisted as a 'duty' among non-religious parents and that for many people it is 'a social ethnic tradition, a symbol of respect paid to the family and community'.

27. Dyer, 1991, p42.

28. Golovnev, 1994, p67 (He also notes that this was connected with an idea that souls are associated with four parts of a female body but five parts of a male body).

29. Influences from Judaism could come from at least three sources: 1) the ancient Jewish diaspora communities in cities such as Bukhara: there are biblical references to Jews living in Parthia (Acts 2:9), some of whom might have migrated there from the territories of the Medes, to which Jews from the northern kingdom had been deported by the Assyrians (2 Kings 17:6); 2) the Khazars, whose kingdom at its maximum extent encompassed territories now in the northern Caucasus, the Ukraine (including the Crimea), western

After Atheism

Kazakstan and northeastern Uzbekistan: the Khazars were apparently a Turkic people (perhaps with Ugrian or native Caucasian elements) but many of them were attracted to Judaism through the influence of Jewish refugees, so that by the tenth century AD many Khazar documents were written in Hebrew; 3) indirect Jewish influences via Christianity (and perhaps to some extent Islam) in so far as Jewish practices were absorbed into Christianity and the Old Testament held in honour.

30. Among the Buryats, "an ensemble of representations linked to the souls of the dead . . . was revived by the holocaust of the Second World War. The cult of the dead . . . has become a support for the transmission of traditions and for affirmation of ethnic belonging. Relations with the souls of the dead continue to have a shamanic character, even if they are now maintained directly and without a shaman." (Hamayon, 1998, p54.)

31. Perhaps the two merge into one another in some instances. For example, it was when the Russian anthropologist Vladimir Basilov was investigating "survivals of ancestor worship" in Turkmenistan that he discovered an interesting local legend about the supposed ancestor of that group of Turkmen (Basilov, 1994, p83).

32. Esenova, 1998, p451.

33. Williams 1998, pp 301-2.

5. Who am I?

1. Between the censuses of 1979 and 1989, the Russians increased by 5.6%, as compared with growth rates of 45.5% for the Tajiks, 34% for the Uzbeks and Turkmen, 32.8% for the Kyrgyz, 24.1% for the Kazaks and 24% for the Azerbaijanis. However, the Volga Tatars had increased by only 7.4%, and the Bashkorts by only 5.7%. During this period both Islam and Christianity were being suppressed, so it is difficult to attribute these differences directly to religious influences. Moreover, during this same period population growth rates among the Armenians and Georgians — who are at least nominally Christian — were 11.5% and 11.6% respectively. These are higher than the Tatars and Bashkorts, who are at least nominally Muslim. Even if their adherence to Islam might not be as strong as that of some Tajiks and Uzbeks, it nevertheless appears as if the differences in fertility are often cultural more than religious. The similar population growth rates for Russians, Tatars and Bashkorts suggest that these peoples share similar cultural values regarding family size. This is not surprising in view of the long history of intermarriage between Tatars and Russians (Lewis, 1995a, pp 16-17). Feshbach (1984, pp 64-65) suggests that the more moderate rate of increase of the Kazaks and Azerbaijanis, in comparison with Tajiks and Uzbeks, is possibly linked with the Kazaks' close contact with Slavic settlers and with the relatively high degree of urbanisation in Azerbaijan.

2. Lilya Korchagina, personal communication.

3. Similar values are reflected also in Kyrgyz songs. A Kyrgyz woman estimates that about 20% of Kyrgyz songs are about one's mother, 70% about love and 10% about both these themes together.

4. According to Nurilya Shakhanova (personal communication), to some extent a traditional cultural emphasis on fertility among the Kazaks and other nomadic peoples could be reflected by the symbolism used in certain contexts, such as triangles on the edges of the tent at a wedding; sometimes the pole placed at the centre of the tent and ring at the top of the tent could be regarded as symbolising the male and female genitals respectively.

5. A paper of mine on this topic (Lewis, 1997b) was presented at an academic conference in Glazov, Udmurtia, in 1997 but I deliberately read it in the Udmurt language to show respect for the local people: a written translation into Russian was handed out for those who did not understand Udmurt. Later a Mari professor who read the Russian translation commented to me that he liked my paper because I had expressed truths which he himself was afraid to articulate in public.

6. Lewis, 1997b, pp 60-61.

7. Vorontsov 1997, p126 The figures cited are for those who are not from nationally mixed families. For those from mixed backgrounds the corresponding percentage falls to 5% among Udmurts for the question on being humiliated by other ethnic groups (compared with 5% of mixed Russians and none of the mixed Tatars) but for the question about representatives of one's ethnic group not esteeming themselves the percentage is still 25% among mixed Udmurts as compared with 14% of the mixed Russians and none of the mixed Tatars.

8. Lewis, 1997b, pp 60-66.

9. Dewdney, 1993, p217.

10. Khodyreva, 1992; Vladykin & Panova, 1994, p113.

11. Popova, 1997, p7. She does not state where the mullah comes from but he presumably is of Tatar nationality: in the principal districts of northern Udmurtia containing Beserman settlements, the numbers of Tatars recorded by the 1979 census were: 2,634 in the Yukamenskiy district, 337 in the Yar district and 4,134 in the Balezino district.

12. Moss, 1991, pp 240-41; Roshchin 1994, p17.

Notes

13. Moss, 1991, p243, citing Yuri Belov, 'Syashchenniki v lageryakh', in *Posev*, 1980, No. 5, pp 26-28.

14. A somewhat modified version of this same kind of thinking can be found among some Protestant Christians who look beyond particular human beings or government institutions (such as the secret police) to the unseen spiritual powers of darkness: this viewpoint sees the human agents as being unwittingly — even perhaps unwillingly — manipulated and influenced by demonic powers. However, they also recognise that such powers of darkness are also active in many other spheres — including members of the church itself.

15. Language of this kind is commonly found among Russian Protestants and is based on biblical precedents but appears to be less frequently used nowadays among Western Protestants.

16. Theologically they have biblical grounds for such a position: the 'classic' New Testament text is 2 Corinthians 6:14-7:1 but the same concept is also found in the Old Testament in Ezra chapter 10. Old Testament references to marriage within the community of Israel, however, could be interpreted as marriage either within one's own religion or within one's own ethnic group. It is nevertheless significant that the Old Testament also contains the book of Ruth, which gives special honour to a Moabite woman who had married an Israelite man but had also accepted the Jewish faith. She not only became the great-grandmother of Israel's great king David but is also one the few women — two of whom were known to be Gentiles — who in the New Testament are specifically mentioned as being ancestors of the Messiah (Matthew 1:5).

17. Altan, 1988, p280.

18. Fisher, 1978, pp 169-71; Sheehy & Nahaylo, 1980, p8; Nekrich, 1978, pp 23-34; Williams, 1998, pp 294-95.

19. Karklins, 1986, p157 This might seem at variance with the inclusion of Kazakstan among those republics of the former USSR with 'the highest proportion of nationally mixed families' (Bromlei, 1990, p57), but statistics at the republic level, including intermarriage between Russians and Ukrainians (for example), can obscure the behaviour of local ethnic groups such as the Kazaks (Karklins, 1986, pp 158-59).

20. Karklins, 1986, p160; Dyer (1991, pp 40-42) reports that Dungan mixed marriages were mainly of Dungan men marrying women of other nationalities, including perhaps Russians or Germans, but women who wanted to marry a Russian or non-Dungan could face stiff opposition from the Dungan community.

21. Of course, to be officially defined as Tatar for passport purposes does mean that they have at least some kind of Tatar ancestry.

22. See Lewis, 1997a, pp 231-35 for statistics on the geographical distribution of ethnic groups within Tatarstan. The material in this section was originally presented in that article too.

23. Karklins, 1986, p165.

24. Komarova, 1987. (However, one problem in interpreting these figures concerns the social acceptability among each ethnic group of finding a spouse through advertisements. It is possible that Russians view this as a 'last resort' and are therefore less particular about their choice of preferred partner than are Tatars whose cultural backgrounds might, to some extent, make arranged marriages more acceptable. Therefore caution needs to be exercised in the use of such statistics as indicating anything about the strength of a national consciousness among Tatar men and women.).

25. Figures cited on a BBC Radio 4 programme broadcast at 7.20 pm on 3rd December 1991.

26. Dewdney, 1993, p219.

27. Bromlei (1990:, pp 57-58) mentions that this was the case in Chuvashia but in the Baltic states Russian nationality was chosen by only about 50% of the offspring of marriages in which one partner was Russian. Elsewhere there were other local cultural preferences.

28. Lewis, 1997a, pp 218-19.

29. Krag & Funch, 1994, p16.

30. Akbarzadeh, 1997a, pp 66-67; Akbarzadeh, 1997b, pp 517-19; Kirimli, 1997, p56.

31. Paul Fryer, personal communication.

32. Hamayon, 1998, pp 52-53.

6. Finding one's roots

1. Lewis, 1992a, p8.
2. Lewis, 1992a, p9.
3. Filatov, 1997, p278.
4. Bukharaev, 1996, p169, quoting Goulnara Baltanova of Kazan University.
5. This town in Bashkortostan has a population of 104,536 but only 9.4% of them are Bashkorts as compared with 36.9% Tatars: a common stereotype depicts Tatars as somewhat more Islamic than Bashkorts.
6. Bennigsen & Wimbush 1985a, pp 21-24. However, local people tended not to subscribe to this categorisation (Akiner, 1996, p114).
7. Bennigsen & Wimbush, 1985b, pp 51, 61, 67-69. A survey in 1972 also indicated that slightly over a

third of the adepts had chosen a particular spiritual master by 'inheritance' as a family tradition, but approximately another third had made a personal choice themselves (ibid., p59).

8. Ro'i, 1996, p163; Akcali, 1998, pp 268-269.
9. Rohlich, 1986, p167.
10. Filatov, 1997, p276.
11. One example is the theory advocated by Karimullin (1995) who suggests that there might be parallels between the Turkic languages and some of the indigenous languages of America, such as Dakota and Maya. Such theories are probably popular partly because they serve to enhance one's feelings about the importance of one's ancestral roots.
12. Rakhmat Rakhimov, personal communication, 1990.
13. Shamiladze, 1990, p4.
14. This news was reported by a member of the audience who attended a conference on 'Society and History in Soviet Caucasia' in July 1990 at the School of Oriental and African Studies, University of London. I did not obtain his name but he said that this story had been circulated in the Georgian unofficial press and had also been mentioned on Georgian television.
15. Carrere d'Encausse, 1990, p263.
16. Baibosynov & Mustafina, 1989, pp 70-79. In 1990 Vladimir Basilov (personal communication) remarked to me that from isolated reports it seems there were probably about a dozen practising shamans at that time in Kazakstan but it was relatively rare on account of the long history of official repression of religion. However, it was far more common in Uzbekistan and Tajikistan.
17. Hamayon, 1998, p58.
18. Barkmann, 1997, pp 69-70; Hamayon 1998, pp 58-59.
19. Vyatkina, 1964, pp 226-7; Hamayon 1998, pp 53-54, 58.
20. Humphrey, 1989, p18.
21. Shundrin, 1993a, p17.
22. Information derived principally from an interview with Yevgeny Aronson in 1991.
23. This account was related to me by Julia Borisenkova.
24. Shnirelman, 1997; Sergei Arutuniov, personal communication; Krindach 1994, p21.
25. de Cordier, 1997, pp 591, 599; Filatov & Shchipkov (1995, p243) write that 5% to 7% of the inhabitants of the Mari-El republic are 'pure' pagans, 60% of dual faith (both visiting sacred groves and attending church), and 30% (mostly Russians) are Orthodox Christians. However, up to 90% of the 200,000 Mari diaspora in Bashkortostan, Tatarstan and Udmurtia are 'pure' pagans.
26. Filatov & Shchipkov (1995, p237) mention a Mordvin woman who each night "repeated in higgledy-piggledy fashion" both psalms in Slavonic and Erzya prayers to the pagan god Ineshkipaz.
27. Filatov & Shchipkov, 1977, p178.
28. Razin, 1997, pp 46-52.
29. Filatov & Shchipkov, 1977, p179.
30. Razin, 1997, p49.
31. Filatov & Shchipkov 1977, p181.
32. Tatyana Vladykina, personal communication.
33. Lewis, 1994a, p22.
34. Uzzell & Filatov, 1997.
35. I could not say that we had already learned of his status through Korstin.
36. Lewis, 1994a, pp 22-23.
37. To try to counteract this possible concealment of rites by older people and village leaders, I took a different strategy in the village of Lombovozh, where I made sure that my interviews were with ordinary people whom I met in the course of my visits there — building up relationships through visiting almost every year as well as helping people through forms of humanitarian aid.
38. Chichlo, 1985, pp 166-81; Prokofyeva et al., 1964, p537; Vasilevich & Smolyak 1964, p649.
39. According to Likhachev (1990, p434), "it is now clear that the Old Believers were correct, so that . . . the Nikonites had to fake certain documents."
40. Stricker 1990, pp 26-28.

7. The Orthodox kaleidoscope

1. White & McAllister, 1997, pp 239-241.
2. The flames could be interpreted as symbolising the Holy Spirit because in Acts 2:3 the coming of the Holy Spirit was accompanied by a vision in which the disciples 'saw what seemed to be tongues of fire that separated and came to rest on each of them'.

Notes

3. In September 1998 I heard that another architect by the name of Morozov had appeared on the scene towards the end of 1997 or in the early part of 1998 and had submitted other plans for this project. The mayor of Yekaterinburg had not yet decided between the two architects by the time of writing.
4. Agence France Presse, Jul. 22, 1999, cited on the 'Russia Today' web site (http://www.russiatoday.com/news.php3?id=80730&text).
5. In an interview with Dobrynina (1998), Father Men's brother, Pavel Men, also expressed the opinion that the Yekaterinburg incident was not an isolated one.
6. *Frontier*, No. 4, 1998, p9: 'Auto-da-fé in Yekaterinburg' (Oxford, Keston Institute).
7. Smirnov, 1993.
8. Ellis, 1994, pp 1-2.
9. A contrast is to some extent provided by a comment from an employee of the government's department for religious affairs in another city of the Russian Federation, who told me that he respected the Baptists because they lived up to their own moral standards. He this case he was referring not so much to financial practices but to what he knew of certain senior priests in the Orthodox church whose celibacy as monks meant that they had found sexual outlets in homosexual practices.
10. Walters, 1993, pp 20-21.
11. Vladimir Domrachev is now known as Vitaly rather than Vladimir because, in accordance with Orthodox tradition, he had taken a new name on becoming a monk.
12. In the Communist period Dudko was known for the frank way in which he taught his congregation and was willing to address topics normally regarded as 'taboo'. He therefore met opposition from the religious and secular authorities (Ellis, 1986, pp 309-315, 422-426, 430-438).
13. Lewis, 1993b, pp 19-20.
14. Lewis, 1995c, pp 18-20.
15. Russia Today report (published on their web site), Updated Wednesday, April 15, 1998, citing Reuters.
16. The first child had been born before the mother had become a Christian.
17. Babasyan 1995, pp 18-19.

8. Free market religion

1. Lewis, 1995d, p8.
2. *Glavniy prospekt*, 5th-11th January 1995; *Oblastnaya Gazeta*, 23rd. December 1994; *Vecherniy Yekaterinburg*, 20th January 1995; *Na Smenu*, 28th December 1994; *Vecherniy Yekaterinburg*, 10th January 1995; *Yekaterinburgskaya Nedelya*, 13th. January 1995.
3. Lewis, 1994b, p12.
4. Lewis, 1997a, p221.
5. Zolotov, 1998.
6. Bukharaev, 1996, p172.
7. *Impact International*, February 1998.
8. Bukharaev, 1996, p172; Shundrin, 1993b, p17.
9. Sheikh Ravil Gainutdin of the central mosque in Moscow has also complained about the emergence of sects within Islam (Shundrin, 1993b, p17). Nowadays in Central Asia and Dagestan there are conflicts among Muslims because of the influence of more 'fundamentalist' Muslims (called 'Wahhabis') who reject as unorthodox local forms of folk Islam such as the veneration of saints.
10. One example I have heard about concerned a visiting Swedish missionary who in Irkutsk had openly denounced the Orthodox church as "a creation of Satan." It is not surprising that there was an Orthodox backlash: the local Baptists were no longer allowed to continue their prison visitation ministry and an Orthodox priest confiscated all the literature given to the prisoners by the Baptists, apart from the Bibles which the prisoners then had to keep in a special Orthodox devotional room.
11. White & McAllister (1997, p242) give statistics whereby 45% of respondents in 1993 but 75% in 1996 identified themselves as Orthodox. In the same period the corresponding percentages for 'atheist' fell from 42% to 17%, while 'other affiliation' rose from 3% to 5%. However, those regularly attending Orthodox churches and who are more 'committed' are probably only about 1% or 2% of the total population.
12. See Lewis, 1997a, pp 232-4 for population statistics and details on the ethnic composition of the various towns and districts of Tatarstan.
13. Lewis, 1995b, pp 41-2.
14. Lewis, 1997a, p226.
15. Statistics based on the total of 13,580 recognised religious communities in Russia in 1995 show that 55% were Orthodox, 20.1% Protestant (including 8.2% Baptist, 3.1% Pentecostal, 2.0% Seventh Day Adventists, 1.3% Lutheran and 1.4% Jehovah's Witnesses), 19.9% Muslim, 1.8% Old Believers, 1.3% Roman

After Atheism

Catholic, 0.7% Buddhist and 0.5% Jewish (Krindatch 1996, p30). However, Orthodox churches are more dependent upon having their own permanent building whereas the numerically smaller Protestant churches sometimes rent premises used for other purposes on weekdays: therefore the number of their congregations is not necessarily proportionate to their share of the population. An informed estimate of the number of Protestants in Tatarstan in 1999 is about 4,860 people among 50 churches, of all Protestant denominations. This means that Protestant Christians make up only 0.1% of the population of Tatarstan.

16. Lewis, 1994c.
17. Lewis, 1994c, p18.
18. Nikolai Shalapovskiy, personal communication. (I attach some credence to Nikolai's statistics because he was the former editor of the newspaper 'Protestant' and has also worked for the Russian Bible Society and as a consultant for a number of Christian organisations.)

9. Human Rights

1. Lewis, 1998, p194.
2. Corley, 1995a, pp 6-7.
3. Krindatch, 1996, p31.
4. Shundrin, 1992, pp 14-16.
5. Lewis, 1992b, p17.
6. Momen, 1991, pp 281-293.
7. Shundrin, 1994, p22; Krindatch (1996, p31) states that the Baha'is "now number some 40 communities."
8. Lewis, 1991b, p13.
9. Dmitriuk, 1990.
10. As I only have Lena's own account, it is impossible to say whether the woman was actually cured or if she now no longer admits to Lena that she has headaches.
11. Lewis, 1991b, p12.
12. *Open Doors* magazine, November 1990, p11.
13. Ellis, 1996, pp 170-90.
14. I am grateful to Lawrence Uzzell of the Keston Institute for his help in supplying me with the detailed text of the law already translated into English.
15. Jawad & Tadjbakhsh, 1995, pp 12-13.
16. See the news reports published by the Keston Institute in their magazine *Frontier*, No. 3, 1998, p3, & 1999, No. 5, pp 2-3.

10. Compassion

1. Lewis, 1993f, p18.
2. Lewis, 1993e, p17; Corley, 1998, p308.
3. Lewis, 1993g, pp 21-22; Corley, 1998, p307.
4. Lewis, 1996a, pp 8-9.
5. US Department of State Armenia Country Report on Human Rights Practices for 1997, Released by the Bureau of Democracy, Human Rights, and Labor, January 30, 1998 (available on the internet at: www.ayf.org/sections/ info/hr_97_armenia.html). Details of electoral abuses in the September 1996 Presidential elections are reported by the International Helsinki Federation for Human Rights in their 1997 annual report (available on the internet at: www.ihf-hr.org/ ar97arm.htm).
6. In the earlier stages of the conflict there were cases of Azerbaijanis rescuing Armenians and also of Armenians rescuing Azerbaijanis. (These are reported in personal testimonies known to me and also mentioned, for example, in the BBC Survey of World Broadcasts, 30th March 1988 and *The Observer*, 27th March 1988.)
7. US Department of State Armenia Country Report on Human Rights Practices for 1997, Released by the Bureau of Democracy, Human Rights, and Labor, January 30, 1998; Amnesty International report *Armenia: Comments on the Initial Report Submitted to the United Nations Human Rights Committee* (Amnesty International-Report-EUR 54/05/98), September 1998. (This is available on the internet at www.amnesty.org/ailib/aipub/1998/EUR/ 45400598.htm.)
8. Details are given in the speeches by Baroness Cox and the Earl of Shannon in a debate in the House of Lords on Wednesday 17th March 1999 as recorded in the House of Lords official report, Vol. 598, No. 54, columns 800-4.
9. Armenian claims to the territory date back to antiquity but the contemporary problem is also a reflection of different stages in the expansion of the Russian empire. In 1813 the province containing

Notes

Nagorno Karabakh was annexed by the Russian empire and incorporated into what later became Azerbaijan, whereas other Armenian territories were not ceded to Russia by Persia until 1828. From then on Nagorno Karabakh was administratively separated from other Armenian-speaking areas. However, on 1st December 1920 the president of the revolutionary committee of Azerbaijan had publicly announced that Nagorno Karabakh and other Armenian-speaking territories within Azerbaijan would be given to Armenia. This promise was confirmed on 3rd July 1921 by the plenary session of the Caucasian bureau of Soviet Russia's Communist Party Central Committee. Two days later, however, on 5th July 1921, the same plenary session revoked their decision and announced that the area would become an 'autonomous region' under the jurisdiction of Azerbaijan. The reasons for this reversal of policy within 48 hours have never been publicly revealed (Libaridian, 1988, pp 5-7, 33-37; Akiner, 1986, p111).

10. *BBC Summary of World Broadcasts*, 12 February 1994.
11. *Covcas Bulletin*, 31 May 1995, page 7, cited by the UNHCR report.
12. Kutzian, 1995.
13. Background paper on refugees and asylum seekers from Armenia issued in August 1995 by the United Nations High Commission for Refugees.
14. Kutzian, 1995.
15. The article is available on the internet at http://rbhatnagar.ececs.uc.edu:8080/alt_hindu/1995_May_1/msg00028.html. Sarkisyan continues by writing, "I don't see any difference between terrorist acts against journalists last year and nowadays against religious communities..."
16. Higgins, 1995.
17. Corley, 1995b, pp 6-7; Henderson, 1995; Corley, 1999, p13. The attacks are also mentioned in passing in a report by Human Rights watch: www.igc.org/hrw/ worldreport99/europe/armenia.html.
18. Corley 1995b, p7; also an anonymous article entitled 'Fight against prejudice' in Keston Institute's magazine *Frontier*, October-December 1994, pp 14-15 and a web site entitled Persecution of nonapostolic religious organisations in Armenia. Part 5 at http://rbhatnagar. ececs.uc.edu:8080/alt_hindu/1995_May_1/msg00027.html. (The same document also refers to other Krishna devotees being kept in prison who would "probably ... be sent to the Karabakh war".) Moreover, the 'Background paper on refugees and asylum seekers from Armenia' issued in August 1995 by the United Nations High Commission for Refugees stated: "Despite repeated official assurances to the contrary, illegal conscription methods continue in Armenia, including indiscriminate manhunts in the streets of the capital for youths eligible to do military service (*Covcas Bulletin*, 9 November 1994). In many instances during 1994 and in 1995, military recruitment commissioners interfered with the right to privacy of citizens ... Recruitment personnel visited the homes of draft-age men and often threatened or detained the occupants or inflicted material damage to them. Able-bodied males were seized at market places, bus and railway stations and other public places (US Department of State, Country Reports on Human Rights Practices 1995)." The same report from the UNHCR further cites an article of 15th April 1995 by the Noyan-Tapan news agency of Yerevan, which quoted the Chairman of the military draft commission, Vazgen Sarkisyan, as saying that the names of all "deserters" are known, and that they would be "blacklisted" and eventually prosecuted by the courts. "Yet," the UNHCR report continues, "Armenian officials deny reports of forced conscription to boost the Karabakh war effort, and men of military age are stopped from flying out of Yerevan unless they can prove that they have completed their military service (Reuters, 17 May 1995)."
19. Ohanjanian 1998, p11; Amnesty International report EUR 54/01/98, January 1998 entitled Armenia: Summary of Amnesty International's Concerns. (This is available on the internet at www.amnesty.org/ailib/aipub/1998/EUR/45400198.htm.)
20. See Jawad & Tadjbakhsh, 1995.
21. Williams, 1998, p309

11. Miracles

1. Lewis, 1993c, p26.
2. His case is reminiscent of those who lose consciousness or "rest in the Spirit" while receiving ministry in some charismatic churches — a phenomenon which is thought to have a biblical parallel in Mark 9:26-27 (see also Daniel 10:7-11; Revelation 1:17).
3. 1 Kings 17:7-16; 2 Kings 4:1-7.
4. 'Pochemu ikona "plachet"?', *Izvestiya Udmurtskoy Respubliki*, 6 (640), 17 January 1995, p1.
5. Lewis, 1989, pp 21-68, 203-234.
6. Playfair, 1985, pp 240-241.
7. Czaplicka, 1914, pp 309-321.
8. Gardner, 1983, p1932.

9. Gardner, 1983 & 1986.
10. Lewis, 1989, pp 221-228.
11. Chung & Enzinger, 1976; Allen, 1977.
12. The x-rays showed the tibia but not the fibula. Unfortunately I did not manage to obtain specialist medical details about the case so my information was based primarily on the details supplied by Leonid's grandmother. However, when discussing what I knew about the case with a British medical doctor, Andrew Taylor, I learned that it is not uncommon for there to be a difference in the growth rates of a child's legs which later balances itself out. What I know of Leonid's condition fits with it probably being talipes — a distortion in the foot caused by a relatively strong muscle — which can often be corrected by physiotherapy and perhaps by an operation.
13. Lewis, 1995b, p41.
14. Lewis, 1993d, p36.
15. Gardner, 1983, p1932.
16. Lewis, 1989, pp 129-61; 1990, pp 126-33.

12. Facing the Future

1. Gohar Marikian, personal communication.
2. Jesus taught that it is necessary to 'change and become like little children' not only to enter the kingdom of heaven but also to progress in one's relationship with God (Matthew 18:3-4). I wonder if this principle applies not only to a child-like faith but also to the ability to hear God's voice.
3. In this instance, it is unclear whether the daughter was responding to a 'premonition' or something like 'telepathy': did she write the letter before her father got drunk or at the same time? There is also an ambiguity in this instance about whether or not the father was prone to drinking and therefore the daughter was reflecting her awareness of this.
4. Hoyle, 1983, pp 212-3. Hoyle argues that 'nature is very parsimonious, in the sense that where possibilities exist they seem always to be used'. Therefore 'if the familiar past-to-future time-sense were to lie at the root of biology, living matter would like other physical systems be carried down to disintegration and collapse. Because this does not happen, one must conclude, it seems to me, that biological systems are able in some way to utilize the opposite time-sense in which radiation propagates from future to past'. In this way Hoyle attempts to explain many observable phenomena (such as spiders' webs) which are not explicable in terms of chance evolution but only by an idea that there was an intelligent design from the beginning which knew the final result and how it would function. An unanswered question remains about the source and nature of that intelligence behind the design of the universe.
5. Lewis, 1990, pp 109-120.
6. In further discussion he clarified that it was definitely not 'idle' but 'idol'.
7. In the Bible, this apparently happened to the prophet Daniel (Daniel 10:7-9) and to the apostle John (Revelation 1:17). For further discussion of such phenomena, see Lewis, 1989, pp 181-2, 357-8.
8. Ring, 1984, pp 193-219; Ritchie, 1978, pp 71, 119-21.
9. Grey, 1985, pp 96-7; Ritchie, 1978, pp 48-55.
10. Moody 1975, pp 15-80.
11. Lewis, 1989, pp 360-1. One difference is that the book of Revelation (chapters 2 and 3) contains assessments of the spiritual lives of other people — that is, of churches — rather than that of John himself, but a modern parallel to this is provided by the case of a woman in northern Thailand who had a vision of Jesus but information was also divulged to her about the sins of others in her village (Gardner, 1986:138).
12. Roxburgh, 1991, pp 10ff.
13. Revelation 8: 10-11, New International Version. (A footnote adds the explanatory comment that the name 'Wormwood' in Greek also meant 'Bitterness'.).
14. Shirendyb, 1968.
15. Quoted in Lambert, 1975, pp 63-64.
16. *Guardian*, 9th March 1988; *Times*, 12th March 1988.
17. Hay, 1982, 1987, pp 120-34
18. Kaariainen & Furman, 1997.

Bibliography

Akbarzadeh, Shahram, 'A note on shifting identities in the Ferghana valley', *Central Asian Survey*, 16: 1, March 1997, 1997a, pp 65-68.

Akbarzadeh, Shahram, 'The political shape of Central Asia', *Central Asian Survey*, 16:4, December 1997, 1997b, pp 517-42.

Akcali, Pinar, 'Islam as a "common bond" in Central Asia: Islamic Renaissance Party and the Afghan mujahidin', *Central Asian Survey*, 17: 2, June 1998, 1998, pp 267-84.

Akiner, Shirin, *Islamic Peoples of the Soviet Union*, revised edition, London-New York-Sydney-Melbourne: Routledge & Kegan Paul, 1983/1986.

Akiner, Shirin, 'Islam, the state and ethnicity in Central Asia in historical perspective', *Religion, State and Society*, 24:2/3, 1996, pp 91-132.

Algar, Hamid, 'Shaykh Zaynullah Rasulev: The Last Great Naqshbandi Shaykh of the Volga-Urals Region', in Jo-Ann Gross (ed.), *Muslims in Central Asia: Expressions of Identity and Change*, Durham, USA-London, Duke University Press, 1992.

Allen, P. W., 'The fibromatoses: a clinicopathologic classification based on 140 cases', *American Journal of Surgical Pathology*, 1, 1977, pp 255-70, 305-21.

Allworth, Edward (ed.), *Tatars of the Crimea*, Durham, USA-London, Duke University Press, 1988.

Altan, Mübeyyin Batu, 'Structures: the importance of family — a personal memoir', in Edward Allworth (ed.), *Tatars of the Crimea*, Durham, USA & London, Duke University Press, 1988.

Atkin, Muriel, 'Religious, national and other identities in Central Asia', in Jo-Ann Gross (ed.), *Muslims in Central Asia: Expressions of Identity and Change*, Durham, USA-London, Duke University Press, 1992.

Babasyan, Nataliya, 'Praise Him with cymbals', *Frontier*, March-May 1995, pp 18-19.

Baibosynov, K. & Mustafina, R., 'Novye svedeniya o Kazakhskikh shamankakh', *Novoe v Ethnografii*, 1, pp 70-79, 1989.

Barkmann, Udo B., 'The revival of lamaism in Mongolia', *Central Asian Survey*, 16: 1, March 1997, 1997, pp 69-79.

Basilov, V.N., 'Legend: to believe or not to believe?', in Tamara Dragadze (ed.), *Kinship and Marriage in the Soviet Union*, London-Boston-Melbourne-Henley, Routledge & Kegan Paul, 1994.

Beliaev, Igor, 1990, 'Islam and Politics', in Martha B. Olcott (ed.), *The Soviet Multinational State*, New York & London, M. E. Sharpe.

Bennigsen, Alexandre & Lemercier-Quelquejay, C., '"Official" Islam in the Soviet Union', *Religion in Communist Lands*, 7, 1979, pp 148-59.

Bennigsen, Alexandre & Wimbush, S. Enders, *Muslims of the Soviet Empire*, London, Hurst, 1985a.

Bennigsen, Alexandre & Wimbush, S. Enders, *Mystics and Commissars: Sufism in the Soviet Union*, London, Hurst, 1985b.

Binns, Christopher A. P., 'The changing face of power: revolution and accommodation in the development of the Soviet ceremonial system', parts I & II, *Man*, 14, 1979, pp 585-606; *Man*, 15, 1980, pp 170-187.

Boyes, Roger, 'Witch guide to the mystic East', *The Times*, Saturday May 30, 1998.

Bromlei, Iu. V., 'Ethnic processes in the USSR', in Martha B. Olcott (ed.), *The Soviet Multinational State*, New York & London, M. E. Sharpe, 1990.

Buckley, Cynthia, 'Suicide in post-Soviet Kazakhstan: role stress, age and gender', *Central Asian Survey*, 16: 1, March 1997, 1997, pp 45-52.

Bukharaev, Ravil, 'Islam in Russia: crisis of leadership', *Religion, State and Society*, 24, nos.2/3, Sept. 1996, pp 167-182.

Burridge, Kenelm, *New Heaven, New Earth*, Oxford, Blackwell, 1969.

Carrere d'Encausse, Helene, *The End of the Soviet Empire*, New York, Basic Books, 1990.

Chichlo, Boris, 'The cult of the bear', *Religion in Communist Lands*, 13: 2, 1985, pp 166-181.

Chung, E. B. & Enzinger, F. M., 'Infantile fibrosarcoma', *Cancer*, 38, 1976, pp 729-739.
Chylinski, Ewa A., 'Ritualism of family life in Soviet Central Asia: the sunnat (circumcision)', in Shirin Akiner (ed.), *Cultural Change and Continuity in Central Asia*, London & New York, Kegan Paul International in association with the Central Asia Research Forum, School of Oriental and African Studies, University of London, 1991.
Corley, Felix, 'AUM cult fights back', *Frontier*, June-August 1995, 1995a, pp 6-7.
Corley, Felix, 'Might, not right', *Frontier*, September-October 1995, 1995b, pp 6-7.
Corley, Felix, 'The Armenian church under the Soviet and independent regimes', *Religion, State and Society*, 26: 3/4, 1998, pp 291-355.
Corley, Felix, 'Face to face with the government and church', *Armenian Forum*, 1: 4, 1999, pp 1-18.
Crisp, Simon, 'Census and sociology: evaluating the language situation in Soviet Central Asia', in Shirin Akiner (ed.), *Cultural Change and Continuity in Central Asia*, London & New York, Kegan Paul International in association with the Central Asia Research Forum, School of Oriental and African Studies, University of London, 1991.
Czaplicka, M. A., *Aboriginal Siberia*, Oxford, Clarendon Press, 1914.
Czaplicka, M. A., *Collected Works*, ed. David Collins, London, Curzon Press, 4 vols., 1998.

de Cordier, Bruno, 'The Finno-Ugric peoples of central Russia: opportunities for emancipation or condemned to assimilation?', *Central Asian Survey*, 16: 4, December 1997, pp 587-609.
Dewdney, J.C., 'The Turkic peoples of the USSR', in Margaret Bainbridge (ed.), *The Turkic Peoples of the World*, London & New York, Kegan Paul International, 1993.
Dmitriuk, Mikhail, 'Acting with the force of Nature', *Soviet Weekly*, 18th January 1990, p5.
Dobrynina, Svetlana, 'Inkvizitsiya: 98', *Podrobnosty*, 11th June 1998.
Douglas, Mary, *Purity and Danger*, London, Routledge & Kegan Paul, 1966.
Dyer, Svetlana Rimsky-Korsakoff, *Soviet Dungans in 1985: Birthdays, Weddings, Funerals and Kolkhoz Life*, Taipei, Center for Chinese Studies, 1991.

Ellis, Jane, *The Russian Orthodox Church*, Bloomington & Indianapolis, Indiana University Press, 1986.
Ellis, Jane, 'Priest in exile', *Frontier*, March-April 1994, pp 1-2.
Ellis, Jane, *The Russian Orthodox Church: Triumphalism and Defensiveness*, Basingstoke, Macmillan Press in association with St Antony's College, Oxford, 1996.
Esenova, Saulash, '"Tribalism" and identity in contemporary circumstances: the case of Kazakstan', *Central Asian Survey*, 17: 3, September 1998, pp 443-62.
Evans-Pritchard, E. E., *Social Anthropology*, London, Cohen & West, 1951.
Evans-Pritchard, E. E., *Theories of Primitive Religion*, Oxford, Clarendon Press, 1965.

Feldman, Walter, 'Interpreting the Poetry of Mäkhtumquli', in Jo-Ann Gross (ed.), *Muslims in Central Asia: Expressions of Identity and Change*, Durham & London, Duke University Press, 1992.
Feshbach, Murray, 'Trends in the Soviet Muslim population: demographic aspects', in Yaacov Ro'i (ed.), *The USSR and the Muslim World*, London, George Allen & Unwin, 1984.
Filatov, Sergei & Shchipkov, Aleksandr, 'Religious developments among the Volga nations as a model for the Russian Federation', *Religion, State and Society*, 23: 3, Sept. 1995, pp 233-48.
Filatov, Sergei & Shchipkov, Aleksandr, 'Udmurtia: orthodoxy, paganism, authority', *Religion, State and Society*, 25: 2, June 1997, pp 177-83.
Filatov, Sergei, 'Religion, power and nationhood in Sovereign Bashkortostan', *Religion, State and Society*, 25: 3, Sept. 1997, pp 267-80.
Fisher, Alan W., *The Crimean Tatars*, Stanford, California, Hoover Institution Press, 1978.
Forsyth, James, *A History of the Peoples of Siberia*, Cambridge, Cambridge University Press, 1992.

Gardner, Rex, 'Miracles of healing in Anglo-Celtic Northumbria as recorded by the Venerable Bede and his contemporaries: a reappraisal in the light of 20th-century experience', *British Medical Journal*, 287, 24-31 December 1983, pp 1927-33.
Gardner, Rex, *Healing Miracles: A Doctor Investigates*, London, Darton, Longman & Todd, 1986.
Golovnev, Andrei V., 'From one to seven: numerical symbolism in Khanty culture', *Arctic Anthropology*, 31: 1, 1994, pp 62-71.
Gosudarstvenniy Komitet SSSR po statistike [State Statistical Committee of the USSR], *SSSR v Tsifrakh v 1989 Godu* [The USSR in figures in 1989], Moscow, Finansy i statistika, 1990.
Gosudarstvenniy Komitet RSFSR po statistike [State Statistical Committee of the RSFSR], *RSFSR v Tsifrakh v 1989 Godu* [The RSFSR in figures in 1989], Moscow, Finansy i statistika, 1990.

Bibliography

Gouchinova, Elza-Bair, *Respublika Kalmykia: Model' ethnologicheskogo monitoringa*, Moscow: Academy of Sciences, Institute of Ethnology & Anthropology, 1997.
Grey, Margot, *Return from Death*, London & New York, Arkana, 1985.

Hamayon, Roberte N., 'Shamanism, Buddhism and epic hero-ism: which supports the identity of the post-Soviet Buryats?', *Central Asian Survey*, 17:1, March 1998, pp 51-67.
Hay, David, *Exploring Inner Space*, revised edition, London & Oxford: Mowbray, 1982/1987.
Henderson, Patrick, Report on Armenia available on the internet at: http://rbhatnagar.ececs.uc.edu:8080/alt_hindu/1995_May_1/msg00030.html, 1995.
Higgins, Andrew, 'Where true democracy is the first casualty', *The Independent*, 12th June 1995.
Hostler, Charles Warren, *The Turks of Central Asia*, Westport, Connecticut & London, Praeger, 1993.
Hoyle, Fred, *The Intelligent Universe*, London, Michael Joseph, 1983.
Humphrey, Caroline, *The Herders of Mongun-Taiga*, Manchester, Granada Television Disappearing World booklet, 1989, pp 14-19.
Huttenbach, Henry R., 'The Soviet Koreans: products of Russo-Japanese imperial rivalry', *Central Asian Survey*, 12. No. 1, 1993, pp 59-69.

Jawad, Nassim & Tadjbakhsh, Shahrbanou, *Tajikistan: A Forgotten Civil War*, London, Minority Rights Group, 1995.
Jeziorska, Irena, 'Religious themes in the novels of Chingiz Aitmatov', in Shirin Akiner (ed.), *Cultural Change and Continuity in Central Asia*, London & New York: Kegan Paul International in association with the Central Asia Research Forum, School of Oriental and African Studies, University of London, 1991.

Kaariainen, Kimmo, & Furman, Dmitry, 'Tsena "Religioznoy pyatiletki" ', *Obshchaya gazeta*, 13, 19 March 1997.
Karimullin, Abrar, Prototyurki i indeytsy Ameriki, Moscow, Insan, 1995.
Karklins, Rasma, *Ethnic Relations in the USSR: The Perspective from Below*, Boston, London, Sydney, Allen & Unwin, 1986.
Khan, Paul, *The Secret History of the Mongols*, Boston: Cheng & Tsui Company, 1998.
Khodyreva, Marina, *Besermyan Crez'* [Record & explanatory jacket], Izhevsk, Aprelevka Sound Inc, 1992.
Kirimli, Meryem, 'Uzbekistan in the new world order', *Central Asian Survey*, 16:1, March 1997, pp 53-64.
Komarova, G.A., 'Etnicheskie ustanovki pri vybore brachnovo partnera (po materialam sluzhby brachnykh ob'iavlenii v Ufe i Cheboksarakh) [Ethnic arrangements in the choice of marriage partner — based on materials of the service of marriage advertisements in Ufa and Cheboksary]', *Sovetskaia Etnografia*, 3, Moscow, 1987, pp 80-90, cited in *The Central Asian Newsletter*, 7:1, April 1988, pp 8-9..
Krag, Helen & Funch, Lars, *The North Caucasus: Minorities at a Crossroads*, London, Minority Rights Group, 1994.
Krindach, Aleksei, 'Abkhazia — a profile', *Frontier*, May-June 1994, pp 20-21, 1994.
Krindatch, Alexei D., *Geography of Religions in Russia*, Decatur, Georgia [USA], Glenmary Research Center, 1996.
Kutzian, Marina, e-mail reproduced on the internet at http://rbhatnagar.ececs.uc.edu:8080/alt_hindu/1995_May_1/msg00029.html, 1995.

Lambert, Lance, *Battle for Israel*, Eastbourne, Kingsway Publications, 1975.
Lewis, David C., 'All in good faith', *Nursing Times & Nursing Mirror*, 18th-24th March 1987 = 1987a.
Lewis, D. C., 'Psychic Experiences among Nurses', *Numinis*, Oxford, Alister Hardy Research Centre, 1987b.
Lewis, D. C., *Healing: Fiction, Fantasy or Fact?*, London, Hodder & Stoughton, 1989.
Lewis, D. C., Contributions to *What is the New Age?*, by Michael Cole, Jim Graham, Tony Higton & David Lewis, London, Hodder & Stoughton, 1990.
Lewis, D. C., 'Dreams and paranormal experiences among contemporary Mongolians', *Journal of the Anglo-Mongolian Society*, 13, nos. 1 & 2, 1991 = 1991a, pp 48-55.
Lewis, D. C., 'Superstition in the USSR', *Frontier*, May-June 1991 = 1991b, pp 12-13.
Lewis, D. C., 'Divided we stand', *Frontier*, April-June 1992 = 1992a, pp 8-9.
Lewis, D. C., 'Baha'is — Healing Divisions?', *Frontier*, April-June 1992 = 1992b, p17

Lewis, D. C., *The Unseen Face of Japan*, Tunbridge Wells/Crowborough, Monarch, 1993a.
Lewis, D. C., 'From hippy to priest', *Frontier*, January-March 1993 = 1993b, pp 19-20.
Lewis, D. C., 'God's anointed', *Frontier*, April-June 1993 = 1993c, pp 26-27.
Lewis, D. C., 'A Russian Christian mission of mercy', *Healing and Wholeness*, April/June 1993 = 1993d, pp 36-37.
Lewis, D. C., 'Working partners', *Frontier*, July-August 1993 = 1993e, p17.
Lewis, D. C., 'Forgotten children', *Frontier*, September-October 1993 = 1993f, p18.
Lewis, D. C., 'First to care', *Frontier*, September-October 1993 = 1993g, pp 21-22.
Lewis, D. C., 'Living faith', *Frontier*, March-April 1994 = 1994a, pp 22-23.
Lewis, D. C., 'Rival muftis in Tatarstan', *Frontier*, July-September 1994 = 1994b, p12.
Lewis, D. C., 'Siberia — a check-up', *Christian Herald*, 5th November 1994 = 1994c, p18.
Lewis, D. C., 'Downward trends', *Frontier*, January-February 1995 = 1995a, pp 16-17.
Lewis, D. C., 'New light in a Russian city', *Renewal*, 229, June 1995 = 1995b, pp 40-42.
Lewis, D. C., 'New life among the Russian Orthodox', *Renewal*, 230, July 1995 = 1995c, pp 18-20.
Lewis, D. C., 'Keep Russia Orthodox', *Christian Herald*, 22nd July 1995 = 1995d, p8.
Lewis, D. C., 'No compassion', *Frontier*, January-March 1996 = 1996a, pp 8-9.
Lewis, D. C., 'Religiozniy opyt nekotorykh narodov Sibiri [Religious experience among some peoples of Siberia]', in *Religiya i tserkov' v kulturno-istoricheskom razvitii russkogo severa*, Kirov, Kirov Regional Herzen Library, 2, 1996 = 1996b, pp 140-48.
Lewis, D. C., 'Ethnicity and religion in Tatarstan and the Volga-Ural region', *Central Asian Survey*, 16:2, June 1997 = 1997a, pp 215-36.
Lewis, D. C., 'Natsional'naya gordost' i kulturnye tsennosti Finno-Ugorskikh narodov', *Vestnik Udmurtskogo Universiteta*, 8, 1997, pp 60-69. ['National dignity and cultural values of Finno-Ugric peoples', *The Udmurt University Reporter*, 8, 1997, pp 60-69], 1997b.
Lewis, D. C., 'Spiritualniye opyty: rezultaty issledovaniy i metodologiya', V. A. Korshunkov, I. Yu. Trushkova & A. G. Shurygina (eds.), *Etnokulturnoe nasledie Vytsko-Kamskogo regiona, problemy, poiski, resheniya*, Kirov, Kirov Regional Teacher Training Institute/Kirov Regional Museum of Local History, 1998a.
Lewis, D. C., 'Years of calamity: Yakudoshi observances in urban Japan', in Joy Hendry (ed.), *Interpreting Japanese Society*, London & New York, Routledge, 2nd edn, 1998b.
Libaridian, Gerard J. (ed.), *The Karabagh File*, Cambridge, Massachusetts & Toronto, Zoryan Institute for Contemporary Armenian Research and Documentation, 1988.
Likhachev, D. S., 'Preliminary results of a thousand-year experiment', in Martha B. Olcott (ed.), *The Soviet Multinational State*, New York & London, M. E. Sharpe, 1990.

Momen, 'The Baha'i community of Ashkhabad: its social basis and importance in Baha'i history', in Shirin Akiner (ed.), *Cultural Change and Continuity in Central Asia*, London & New York, Kegan Paul International in association with the Central Asia Research Forum, School of Oriental and African Studies, University of London, 1991.
Moody, Raymond A., *Life after life*, New York, Guideposts, 1975.
Moss, Vladimir, 'The True Orthodox Church of Russia', *Religion in Communist Lands*, 19:2-3, Winter 1991.
Musk, Bill, *The Unseen Face of Islam*, Tunbridge Wells, Monarch, 1989.

Nekrich, Aleksandr M., *The Punished Peoples*, New York, W. W. Norton & Company, 1978.
Nikitina, Galina A., *Narodnaya pedagogika udmurtov*, Izhvesk, Udmurtia Publishers, 1997.
Ohanjanian, Karen, 'Religious freedom versus national security in Nagorno-Karabakh', *Frontier*, 1, 1998, p11.
Olcott, Martha Brill, *The Kazakhs*, Stanford, California, Hoover Institution Press, 1987.
Ostrander, Sheila, & Schroeder, Lynn, *Psi, Psychic Discoveries behind the Iron Curtain*, London, Sphere Books, 1970.

Pika, Aleksandr, 'The spatial-temporal dynamic of violent death among the native peoples of Northern Siberia', *Arctic Anthropology*, 30:2, 1993, pp 61-76.
Pika, A., & Bogoyavlensky, D., 'Yamal peninsula: oil and gas development and problems of demography and health among indigenous populations', *Arctic Anthropology*, 32:2, 1995, pp 61-74.
Pika, Aleksandr I., 'Reproductive attitudes and family planning among the aboriginal peoples of Alaska, Kamchatka, and Chukotka: the results of comparative research', *Arctic Anthropology*, 33:2, 1996, pp 50-61.

Bibliography

Playfair, Guy Lyon, *If This Be Magic*, London, Jonathan Cape, 1985.
Platt, Jennifer, *Realities of Social Research*, Brighton, Sussex University Press, 1976.
Popova, E. V., 'Besermyane', in G. K. Shklyayev (ed.), *O Besermyanakh*, Izhevsk, Academy of Sciences, 1997.
Prokofyeva, E. D., Chernetsov V .N., & Prytkova, N. F., 'The Khants and Mansi', in M. G. Levin & L. P. Potapov (eds.), *The Peoples of Siberia*, Chicago & London, University of Chicago Press, 1964.

Razin, A. A., 'Maly mon udmurt Inmarly vösyas'kis'ko?' [Малы мон удмурт Инмарлы вöсяськисько? Why do I pray to the Udmurt God?], *Vestnik udmurtskogo universiteta* [The Udmurt University Reporter], 8, 1997, pp 46-52.
Ring, Kenneth, *Heading Toward Omega*, New York: William Morrow & Co., 1984.
Ritchie, George, *Return from Tomorrow*, Eastbourne: Kingsway, 1978.
Rogozin, P. I., *Otkuda vsyo eto poyavilos'?*, Rostov on Don: Missiya Probuzhdenie, 1993.
Ro'i, Yaacov, 'Islam in the Soviet Union after the Second World War', *Religion, State and Society*, 24:2/3, Sept. 1996, pp 159-66, 1996.
Rohlich, Azade-Ayse, *The Volga Tatars: A Profile in National Resilience*, Stanford, California: Hoover Institution Press, 1986.
Roshchin, Mikhail, 'Out from underground', *Frontier*, March-April 1994, p17, 1994.
Rowe, Michael, *Russian Resurrection*, London, Marshall Pickering, 1994.
Roxburgh, Angus, *The Second Russian Revolution*, London, BBC Books, 1991.

Sadikov, D., 'Modern Kirghiz', paper given at a conference on languages and scripts of Central Asia held at the School of Oriental and African Studies, University of London, 2nd-3rd April 1990, 1990.
Schmidt, W., *The Origin and Growth of Religion*, London, Methuen, 1931.
Shamiladze, V., 'The ethno-cultural and political situation in Adjara and the problem of its autonomisation (XX century, the period 10th-20th years)', paper presented at a conference on Society and History in Soviet Caucasia, 17th-19th July 1990 at the School of Oriental and African Studies, University of London, 1990.
Sheehy, Ann & Nahaylo, Bohdan, *The Crimean Tatars, Volga Germans and Meskhetians: Soviet Treatment of some national minorities*, London, Minority Rights Group, 1980.
Shirendyb, B., *By-passing Capitalism*, Ulaan Baator: State Publishers, 1968.
Shnirelman, Victor, 'Neo-paganism', paper presented at a conference on Religion in Post-Soviet Russia organised by King's College Research Centre, Cambridge, 29th September-1st October 1997.
Shundrin, Aleksandr, 'They serve Krishna', *Frontier*, April-June 1992, pp 14-16.
Shundrin, Aleksandr, 'Turn of the tide', *Frontier*, April-June 1993 = 1993a, pp 16-18.
Shundrin, Aleksandr, 'Update on Islam', *Frontier*, November-December 1993 = 1993b, p17.
Shundrin, Aleksandr, 'Second lease of life', *Frontier*, May-June 1994, p22.
Smirnov, M., 'Vladyka sluzhit za butilku [The archbishop serves for a bottle]', *Izvestiya*, 17th November 1993.
Southwold, Martin, 'Buddhism and the definition of religion', *Man*, n.s., 13:3, 1978, pp 362-379.
Southwold, Martin, *Buddhism in Life*, Manchester, Manchester University Press, 1983.
Stepanova M.V., Gurvich I.S. & Khramova, V.V., 'The Yukagirs', in M.G. Levin & L.P. Potapov (eds.), *The Peoples of Siberia*, Chicago & London, University of Chicago Press, 1964.
Stricker, Gerd, 'Old Believers in the territory of the Russian Empire', *Religion in Communist Lands*, 18:1, Spring 1990.

Taheri, Amir, *Crescent in a Red Sky: The Future of Islam in the Soviet Union*, London, Melbourne, Auckland, Johannesburg, Hutchinson, 1989.
Tishkov V. A. (ed-in-chief), *Narody Rossii Entsiklopediya* [Encyclopedia of the Peoples of Russia], Moscow, Scientific Publisher 'The Great Russian Encyclopedia', 1994.
Towler, Robert, *Homo Religiosus*, London, Constable, 1974.
Tugolukov, V. A., 'Some aspects of the beliefs of the Tungus (Evenki and Evens)', in V. Diószegi & M. Hoppál (eds.), *Shamanism in Siberia*, Budapest, Akadémiai Kiadó, 1978.
Tylor, Edward B., *Primitive Culture*, 1, London, John Murray, 1873.

Uzzell, Lawrence & Filatov, Sergei, 'The revival of paganism in north-eastern Siberia', *Frontier*, 3, 1997, pp 8-9.

Vasilevich, G. M. & Smolyak, A. V., 'The Evenks', in M.G. Levin & L.P. Potapov (eds.), *The Peoples of Siberia*, Chicago & London, University of Chicago Press, 1964.

Vasilevich, G. M., 'The acquisition of shamanistic ability among the Evenki (Tungus)', in V. Dioszegi (ed.), *Popular Beliefs and Folklore Tradition in Siberia*, Bloomington: Indiana University & The Hague, Mouton, 1968.

Vasiljev, V. I., 'Animistic notions of the Enets and Yenisei Nenets', in V. Diószegi & M. Hoppál (eds.), *Shamanism in Siberia*, Budapest: Akadémiai Kiadó, 1978.

Vdovin, I. S., 'Social foundations of ancestor cult among the Yukagirs, Koryaks and Chukchis', in V. Diószegi & M. Hoppál (eds.), *Shamanism in Siberia*, Budapest, Akadémiai Kiadó, 1978.

Vladykin, V. E. & Panova, E.V., 'Besermanye', in V.A. Tishkov (principal ed.), *Narody Rossii Entsiklopeiya* [Encyclopedia of the Peoples of Russia], Moscow, Bolshaya Rossiyskaya Entsiklopediya, 1994.

Vorontsov, V. S., 1997, 'Opredelenie svoey etnicheskoy prinadlezhnosti podrostkami v natsional'no smeshannykh semyakh', *Vestnik Udmurtskogo Universiteta*, 8, 1997, pp 113-132) ['Determinates of one's ethnic belonging among teenagers in nationally mixed families', *The Udmurt University Reporter*, 8, 1997, pp 113-132]

Vyatkina, K. V., 1964, 'The Buryats', in M.G. Levin & L.P. Potapov (eds), *The Peoples of Siberia*, Chicago & London, University of Chicago Press.

Walters, Philip, 'More questions than answers', *Frontier*, April-June 1993, pp 19-21.

White, Stephen & McAllister, Ian, 'The politics of religion in postcommunist Russia', *Religion, State and Society*, 25:3, Sept. 1997, pp 235-52.

Williams, Brian Glyn, 'The Crimean Tatar exile in Central Asia: a case study in group destruction and survival', *Central Asian Survey*, 17:2, June 1998, pp 285-317.

Zolotov, Andrei, 'Showdown Looms for Tatarstan's Moslems', *Moscow Times*, February 14, 1998.

Index

Abkhaz 141, 148, 157
abortion 99, 198, 272, 274
Adjars 148
adultery 124, 125
Adygei 140
alcohol 23-29, 41, 78, 86, 99, 101, 106, 113, 107, 108, 110, 112-114, 124, 127, 130, 168, 172, 185, 187, 188, 200, 228, 231, 265, 278, 279, 300, 310
amulets 79-81, 233
Andryushkino 31, 34-37, 39-40, 69
ancestors 40, 43, 50, 70, 76, 100-119, 122, 123, 118, 122, 123, 128, 132, 147, 158, 171, 295, 302-306
angels 13-15, 18, 68, 290
Armenia, Armenians 12, 43, 44, 70-71, 239-262, 277, 281, 288-292, 298, 304, 308, 309
astrology 42, 88, 97, 295
atheism 13-17, 19, 28-29, 47, 53, 58, 60, 61, 64, 70, 80, 85, 88, 94, 97, 99, 111, 136, 139, 144, 154, 167, 169, 192, 229, 230, 237, 249, 250, 265, 275, 279, 280, 283, 291, 293-295, 300
Azerbaijan, Azerbaijanis 44, 135, 241, 243, 246, 249, 256, 257, 291, 292, 296, 299, 300, 300, 304, 309

Baha'is 42, 139, 211, 216, 227-231, 259
Bak, Pavel 189, 203, 260-261, 271
Balkars 48, 140
Bashkirs: see Bashkorts.
Bashkorts, Bashkortostan 11, 14, 43, 44, 69, 86, 90, 125, 136, 137, 140, 142, 144-147, 164, 194, 198, 206-208, 224, 247, 271, 299, 300, 304-306
Beserman 132
Berezovo 27, 168, 173
Borisenkovs, Sergei and Julia 157, 213-215, 269, 270
Britain 12, 14, 16, 18, 266-268, 288, 293
'Brotherhood' movements 211, 241-244, 261, 291
Buddhism 14-16, 29, 47, 48, 52, 55, 57-59, 111, 112, 121, 129, 150-154, 176, 223, 224, 293, 295, 300, 302, 308
Buryats 15, 47-48, 121, 140, 150, 151, 224, 304

Caucasus region 41, 43, 44, 48, 50, 119-121, 135, 140, 146, 147, 149, 176, 237, 239, 295, 296, 303, 309
Central Asia 14, 21, 29, 41-44, 50, 99, 118-121, 135, 140, 147-150, 176, 232, 237-239, 251, 273, 295, 299, 300, 303, 304
Chechens, Chechnya 44, 46, 48, 146, 296
Cherkesia 140
Chernobyl 155, 287
Cherskiy 37, 87, 101, 102
children: attitudes towards 24, 26, 37-40, 79-82, 96, 99, 100-103, 119-121, 125, 127, 137, 139, 162, 164, 240, 241, 279, 279, 307, 310
China 16, 47, 48, 57, 87, 88, 120, 288, 303
Christianity 14-16, 19, 40, 43, 47, 50, 64, 65, 69, 85, 90-98, 104, 114, 118, 119, 121, 122, 129, 133-135, 139, 141, 148, 156, 157, 168, 174-206, 208-223, 229, 230, 233, 341-245, 247-274, 281, 282, 286-288, 290-292, 294-297, 300-310
Chukchi, Chukotka 12, 23-28, 30-31, 34, 40, 120, 267, 269, 294, 300-302
Chuvash 46-47, 132, 140, 158, 163, 266, 301, 305
cleanliness (and related concepts) 58, 61-62, 80, 82, 86, 87, 89, 108, 126, 127, 134, 158, 159, 174, 192, 279, 283, 284
coercion 15, 89, 90, 93, 94, 258-260, 296, 297
coincidence 17, 53, 86, 102, 274, 265, 266, 272, 274, 280
Communism 13-18, 32, 36, 53, 61, 86, 90, 91, 94, 122, 131, 133, 136, 137, 139, 140, 148, 151, 167, 168, 169, 172, 177, 185, 193, 197, 202, 205, 215, 218, 225, 228, 233-235, 240, 241, 244, 250, 251, 260, 261, 263, 273, 280, 287, 293, 295, 296, 299, 300, 303, 307, 309
conscience 28, 299, 300
corruption 92, 183-186, 207, 208, 238, 253-255, 261, 281, 288, 292, 295-297
Crimean Tatars 48, 123, 135
curses 76-80, 82, 86, 158, 159, 170, 295, 302

death 24, 26, 27, 31, 52, 54, 55, 57, 58, 61, 63, 64, 70, 73, 75, 76, 78, 82, 87, 88, 95-123, 125, 150, 151, 159, 164, 170, 181, 185, 201, 240, 241, 272, 276, 277, 282-286, 290-292, 296, 300, 302, 303
déjà-vu 60, 62, 63, 65-66, 72
demons/evil spirits 15, 65, 71, 79, 81, 86, 87, 90, 99, 108, 110, 153, 264, 265, 267, 272,

272, 280, 283, 284, 287, 288, 300, 301, 303, 305; *see also* spirits
divination 42, 50, 88, 100, 152, 231, 295
Dolgans 40
Dudko, Dmitry 192, 205, 307
Dungans 120, 303, 305
dreams 13, 17, 18, 20, 51-66, 68, 71-73, 97, 99, 105, 107, 112, 113, 159, 170, 200, 272, 274-278, 280-283, 288, 291, 293, 294, 301

earthquakes 88, 90, 242, 245, 246, 247, 251, 253, 254, 275, 277, 289-292
education 37-40, 50, 52, 54, 60, 77, 86, 88, 125, 129, 130, 149, 154, 155, 187, 233, 236, 265
ethics 16, 17, 28, 94, 199, 200, 233, 239, 256, 273, 281, 283, 290-292, 294, 295-297, 299, 300
Eskimos 30, 31, 40
ethnic identity 17, 34, 128-141, 146-148, 154, 157, 167, 172, 173, 201, 205, 237, 259, 261, 262, 299, 303, 304
Evens 31, 35, 37, 40, 87, 102, 104
Evenki 31, 99, 104, 303
evil eye 79-85, 92, 120, 179, 233, 295
evolution 251, 310

fasting 64, 153, 207, 233, 273, 299
fate 15, 68
feeding the fire 61, 75, 301
fear 13-15, 72-95, 273, 278, 294, 296
fertility 150, 162, 164, 304
Finno-Ugric peoples 15, 30-32, 37-40, 47, 48, 82, 92, 93, 97-99, 129, 157-173, 157-173, 198, 294, 301-304, 306
forgiveness 28, 59, 252, 283, 292

Germans 48, 305
Georgia, Georgians 43, 44, 48, 141, 148, 243, 246, 247, 304, 306
ghosts 68, 86-88, 295, 302
God (as Supreme Being) 14, 15, 17, 53, 54, 55, 59-60, 63-66, 68, 71, 86, 88, 96, 98, 99, 110, 111, 113, 125, 127, 134, 157, 171, 178, 181, 183, 186, 190, 194, 196, 215, 229-231, 242, 250-253, 262, 267, 269, 271, 272, 275, 279-282, 284, 286, 288-293, 297, 299, 301, 302, 310
god/gods (within polytheism) 70, 75-78, 104, 110, 118, 171, 301, 306; *see also* spirits
graves 43, 73, 86-88, 104-106, 113, 114, 116, 117, 119, 142, 175, 184, 302, 303
Greeks 48
Gutiun 240, 241, 244-250, 253-257, 259
Gyumri (Leninakan) 242, 246, 247, 277

Hathaway, David 220, 221, 270, 272
healing 19, 71, 150, 161, 194, 195, 220, 232, 263, 264, 266-274, 292
homosexuality 183, 215, 307
Hui 120
human rights 202, 221, 234-239, 258, 261, 294, 296, 297, 308, 309
humanitarian aid 12, 32, 173, 186-191, 195, 196, 203, 204, 221, 222, 239-241, 244-247, 260, 261, 292, 306

icons 16, 40, 82, 99, 134, 169, 176, 177, 182, 184, 194, 195-197, 263-265
'idols' 75, 77, 104, 171, 173, 281, 289, 301, 310
illness or accidents 13, 50, 71, 75, 78, 79, 80, 82, 88, 150-153, 161, 255, 275-278, 282-285
Ingush 48, 146
interpretations (of spiritual experiences) 13-15, 17-19, 102, 286-292, 301, 306
intuitions 67-68, 99, 272, 274, 276, 284
Islam 13, 14, 29, 41-43, 46-47, 65, 78, 79, 86, 90-92, 103, 115, 117, 119-121, 125, 129, 132, 139, 142-150, 158, 164, 176, 179, 205-208, 216, 221, 224, 227, 229, 234, 238, 251, 262, 273, 279, 293, 295-297, 299-305, 307

Japan 16, 225, 250, 302
Jesus Christ 16, 64, 65, 69, 71, 122, 139, 156, 162, 181, 191, 196, 202, 210, 215, 220, 245, 248, 249, 252, 263, 269, 272, 281, 282, 286, 302, 303, 310
Judaism 121, 127, 130, 131, 154-157, 224, 295, 300, 303-305, 308

Kalmyks, Kalmykia 11, 48, 76, 113, 121, 123, 150, 153, 154, 272, 274, 278
Karabakh 246, 247, 249, 250, 256-260, 290, 291, 308, 309
Karachai 48, 140
Kashpirovsky 233, 272
Kazaks, Kazakstan 12, 41-43, 48, 70, 79-81, 89, 97, 114-118, 120, 121, 123, 135, 139, 149, 150, 186, 227, 294, [delete 297] 299, 300, 302, 304-306
Kazan 13, 46, 91, 126, 136, 137, 142, 143, 145-147, 154-157, 207, 208, 211-217, 224, 225, 229, 231, 266, 268-270, 299
KGB 154, 155, 174, 177, 183, 185, 186, 191-193, 202, 212, 244, 250, 271
Khanty, Khanty-Mansi okrug 11, 26, 30-32, 40, 46, 62, 76-79, 99, 103-106, 119, 121, 129, 130, 132, 139, 168, 170, 294
Kimkyasui 11, 171
Kirov 12, 18, 63
Komi 11, 19, 29, 30, 72, 82-85, 99, 140, 302
Komi-Permyak 303
Koreans 48-50
Koryaks 120, 301

Index

Krishnaism 42, 64, 227, 231, 259, 260, 296, 309
Kumyk 140
Kyrgyz, Kyrgyzstan 42, 44, 123, 135, 238, 297, 300, 304

Lenin 16, 123, 287
Licht im Osten 205
light, spheres of 69, 70, 285, 286, 302
Lombovozh 11, 31-33, 306
Lorino 12, 23-25, 30, 34, 39

Mansi 26-28, 30, 37-39, 46, 69-70, 75, 76-78, 99-101, 103-106, 118, 123, 129, 130, 168-173
Mari, Mari-El Republic 11, 47, 98, 99, 114, 140, 157-163, 165, 167, 304, 306
marriage 28, 46, 50, 76, 91, 98, 99, 107, 108-110, 122, 124-128, 135-139, 148, 164, 177, 178, 247, 273, 282, 295, 302, 304, 305
Marxism 16, 18, 29, 288
mediums 89, 232, 295, 299
Men, Alexander 183, 205, 307
Meskhetian Turks 48, 262
miracles 65, 68, 194, 255, 262-274, 292
Mongolians (Khalkha), Mongolia 14, 19, 41, 47, 48, 52-61, 66-68, 72, 85, 87, 111-113, 150, 151, 275-277, 279, 287, 295, 301, 302
Mordvinians 47, 98, 140, 158, 306
mortality rates 24, 26, 241, 300
Moscow 12, 16, 48, 113, 147, 154-156, 179, 191, 192, 194-197, 205, 209, 216, 219, 221, 225, 227, 251, 276, 281, 282, 287, 294, 307
mosques 42, 43, 90, 142-146, 148, 206-208, 211, 216, 229, 230, 299
Muslims: *see* Islam

nature mysticism 60, 66-67, 231
near-death experiences 97, 282-286
Nenets 30, 31, 40, 62, 77, 79, 104, 131, 132, 301
Nogai 140
nomadic peoples 41, 47, 149, 304
Novosibirsk 12
nurses 14, 18, 65

occultism 213, 225, 231-234, 250, 272, 295
offerings 61, 73, 75, 76, 90, 104-114, 116-118, 160-165, 168, 171-173, 301, 302
Old Believers 16, 133, 135, 174-176, 202, 303, 306, 307
Open Doors 287
Orthodoxy 15, 16, 64, 69, 82, 90-97, 108, 110, 114, 132-134, 154, 161-164, 169, 171, 174, 176-205, 208, 209, 211, 221-224, 232-235, 241, 261, 263-265, 273, 274, 282, 288, 294, 295, 297, 301, 303, 306-308
Osintseva, Tanya 93, 186-190, 204, 205, 271
Ossetians 43, 148

paganism 29, 40, 47, 91, 104-110, 114, 119, 121, 129, 132, 140, 141, 157-173, 231, 295, 301, 303, 306
persecution 17, 130, 131-133, 174, 175, 192, 193, 210, 217, 221, 227, 255, 259, 260, 282, 292, 296, 297, 309
pilgrimage 16, 43, 117, 145, 175, 176, 207, 237
politics 17, 29, 31, 32, 35, 41, 44-48, 50-53, 89-95, 97, 99, 117, 122, 130, 131, 133, 135, 140, 141, 154, 157, 159, 160, 163, 174-176, 185, 193, 206, 207, 225, 227, 233-239, 250, 251, 253-262, 281, 282, 286-288, 290-297
population figures 24, 31, 34, 42-44, 47-49, 77, 98, 131, 132, 142, 154, 168, 210, 217, 300, 304, 305, 308
premonitions 17, 18, 20, 275-280, 284, 293, 310
prisons 133, 189-193, 252, 253, 258, 260, 261, 279, 282, 292, 307, 309
prophecies 70, 170, 221, 281, 282, 286-292
Protestants 14, 94, 98, 133-135, 156, 157, 186, 189, 197-205, 208-223, 227, 233, 234, 306, 241, 248, 259-261, 265, 269-274, 281, 282, 296, 305, 307, 308
psi 15, 280, 284
psychic or 'paranormal' phenomena 13-15, 42, 53, 89, 150, 213, 231-233, 269, 270, 272, 295, 299
purity or pollution (concepts): *see* cleanliness.

Razveyev, Boris 69, 191-194, 263-265
rejection: feelings of 124-126, 129, 130, 190, 272, 273
religion: nature of 14-17, 292-294
'Reuben' 70-71, 281
revelations (from God) 13, 17, 195, 263, 274, 279, 280-282, 291
Roman Catholicism 15, 114, 197, 200, 203, 205, 234, 265, 308
Russians / Russia 14-16, 18, 19, 21, 29, 31, 32, 38-42, 50, 61-64, 72, 76, 78, 82, 85, 88-97, 99, 102, 104, 113, 114, 117, 119, 122-129, 133-141, 155, 158, 159, 165, 171-173, 176-205, 209-227, 231-239, 261, 263-274, 286-288, 294-297, 300-307

Saami 228
sacred places 43, 75, 76, 117, 151, 152, 158-163, 165, 170, 173
sacrifice 61, 106, 114, 116-118, 122, 152, 160, 161, 165, 168, 172, 303

After Atheism

St Petersburg (Leningrad) 12, 48, 147, 154, 219
Sakha (people or republic) 11, 31, 40, 61, 114, 168, 299, 303
salt 80, 82
Sartinya 27
Satan 64, 233, 307
Scripture Gift Mission 205
secret religion 17, 102, 133, 142, 165, 172, 173
shamanism 15, 29, 43, 48, 50-52, 53, 62, 102, 114, 117, 121, 147, 150-152, 159, 168-171, 231, 267, 295, [Delete 297], 299, 301-304, 306
Siberia 14, 15, 18, 23-40, 43, 44, 48, 50, 61-63, 69, 70, 73-79, 87, 90, 92, 100-102, 114, 118, 119, 131, 132, 135, 154, 168-173, 174, 191, 220, 221, 267, 299-304, 306, 307
Sosva 31, 37, 76, 169
speaking 'in tongues' 69, 195, 282
spirits 15, 16, 40, 43, 53-55, 58, 61, 68-70, 73, 75, 79, 81, 85-88, 90, 99-117, 152, 153, 158, 164, 231, 264, 265, 272, 280, 288, 300-303, 305
spiritual experiences 13-19, 52-72, 97, 99-102, 180-182, 191, 192, 275-294, 301, 302, 306, 309, 310
Stalin 16, 48, 123, 135
Stamboltsyan, Khachik 69, 249-255, 257, 283, 284, 292
Sufis 43, 90, 146
suicide 96-99, 302
superstitions 82, 85, 86, 89, 265

Tajiks, Tajikistan 42, 43, 44, 135, 147, 148, 238, 239, 247, 262, 297, 299, 300, 304, 306
Tatars (Volga), Tatarstan 11, 13, 14, 41-46, 79, 90-92, 124-129, 135-140, 142-147, 154, 158, 164, 179, 198, 206-208, 211, 216, 228-230, 238, 239, 272, 279, 282, 300, 302, 304-307

telepathy 13, 14, 20, 53, 267, 283, 284, 310
Teresa, Mother 241, 251
trust -13, 14, 18, 19, 72, 94, 134, 135, 199, 238, 255
Turkmen, Turkmenistan 12, 41, 44, 69, 78-81, 86, 102, 103, 116- 119, 125, 127, 135, 149, 227, 238, 282, 294, 296, 297, 300, 304
Tuvinians 47-48, 150, 152, 153

Udmurts, Udmurtia 11, 47, 75, 82, 92, 93, 97-99, 106-110, 114, 118-120, 129, 132, 140, 141, 158, 163-167, 198, 225, 226, 233, 265, 299, 301, 302, 304, 306
Ufa 86, 137, 144-146, 191-194, 197, 199, 203, 206-208, 218, 224, 247, 263-265
UFOs 42, 64, 69
Uighurs 42, 64-65, 116, 238
urban/rural distinctions 41-42, 49, 50, 127, 128, 149, 150, 304
Uzbeks, Uzbekistan 41-44, 48, 79, 81, 116-119, 145, 238, 239, 261, 262, 296, 297, 304, 306

values 20, 21,29, 58, 99, 118-123, 125-130, 134, 135, 177, 185, 199, 233, 239, 261, 273, 291, 293-297
visions 17, 18, 51, 68-71, 180-182, 201, 243, 277, 278, 281, 285-287, 291, 293, 294, 306, 310

Wahhabis 296, 307
worldviews 13, 62, 85, 90, 121, 123, 171, 265

Yakuts: see Sakha.
Yekaterinburg 12, 64, 133, 142, 174, 177, 179, 180-191, 197, 202-204, 223-227, 231, 232, 260, 271, 272, 287, 307
Yukagirs 31, 34-37, 40, 62, 69, 73-75, 101, 102, 114, 300, 302, 303

Zhirinovsky 131, 157
Zoroastrianism 147, 148, 224, 295